INTEGRATED TECHNOLOGY
FOR PARALLEL IMAGE PROCESSING

INTEGRATED TECHNOLOGY FOR PARALLEL IMAGE PROCESSING

Edited by

S. Levialdi

Department of Mathematics
University of Rome
Rome, Italy

1985

ACADEMIC PRESS, INC.
(Harcourt Brace Jovanovich, Publishers)
London Orlando San Diego New York
Toronto Montreal Sydney Tokyo

ACADEMIC PRESS INC. (LONDON) LTD.
24–28 Oval Road
LONDON NW1 7DX

United States Edition published by
ACADEMIC PRESS, INC.
Orlando, Florida 32887

BRITISH LIBRARY CATALOGUING IN PUBLICATION DATA
Integrated technology for parallel image processing.
1. Image processing
I. Levialdi, S.
621.38'0414 TA1632

ISBN 0-12-444820-8

LIBRARY OF CONGRESS CATALOGING IN PUBLICATION DATA
Main entry under title:

Integrated technology for parallel image processing.

Includes index.
1. Integrated circuits—Very large scale integration—Addresses, essays, lectures. 2. Image processing—Digital techniques—Addresses, essays, lectures. 3. Parallel processing (Electronic computers)—Addresses, essays, lectures. I. Levialdi, S.
TK7874.I548 1985 001.64' 84-18538
ISBN 0-12-444820-8 (alk. paper)
PRINTED IN THE UNITED STATES OF AMERICA

85 86 87 88 9 8 7 6 5 4 3 2 1

Contents

Contributors

Numbers in parentheses indicate the pages on which the authors' contributions begin.

G.B. ADAMS III[1] (133), School of Electrical Engineering, Purdue University, West Lafayette, Indiana 47907

J.D. BECKER (185), Federal Armed Forces University Munich, Faculty of Electrical Engineering, Institute of Physics, D-8014 Neubiberg, West Germany

V. CANTONI (121), Department of Informatics and Systematics, Pavia University, Pavia, Italy

P. DEGANO (1), Computer Science Department, University of Pisa, I-56100 Pisa, Italy

M.J.B. DUFF (153), Department of Physics and Astronomy, University College London, London WC1E 6BT, England

M. FERRETTI (121), Department of Informatics and Systematics, Pavia University, Pavia, Italy

T.J. FOUNTAIN (197), Image Processing Group, University College, London, England

F.A. GERRITSEN[2] (213), Informatics Division, National Aerospace Laboratory NLR, 1059 CM Amsterdam, The Netherlands

R. HARTLEY (101), Center for Automation Research, University of Maryland, College Park, Maryland 20742

H. JAMIESON (57), School of Electrical Engineering, Purdue University, West Lafayette, Indiana 47907

J.T. KUEHN (133), PASM Parallel Processing Laboratory, School of Electrical Engineering, Purdue University, West Lafayette, Indiana 47907

S. LEVIALDI (121), Department of Mathematics, University of Rome, 00185 Rome, Italy

[1]Present address: Research Institute for Advanced Computer Science, NASA Ames Research Center, Moffett Field, California 94035.

[2]Present address: Nederlendse Philips Bedrijven B.V., Medical Systems Division, Software Engineering/Computer Architecture Veenpluis 2, 5684, PC Best, The Netherlands.

F. MALOBERTI (121), Department of Electronics, Pavia University, Pavia, Italy
A. MONKEL (213), Informatics Division, National Aerospace Laboratory NLR, 8316 PR Marknesse, The Netherlands
U. MONTANARI (1), Computer Science Department, University of Pisa, I-56100 Pisa, Italy
GRAHAM R. NUDD (165), Hughes Research Laboratories, Malibu, California 90265
A.P. REEVES (39), School of Electrical Engineering, Cornell University, Ithaca, New York 14853
T.A. RICE (57), School of Electrical Engineering, Purdue University, West Lafayette, Indiana 47907
H.F.A. ROEFS (213), Space Division, National Aerospace Laboratory NLR, 8316 PR Marknesse, The Netherlands
A. ROSENFELD (101), Center for Automation Research, University of Maryland, College Park, Maryland 20742
H.J. SIEGEL (133), PASM Parallel Processing Laboratory, School of Electrical Engineering, Purdue University, West Lafayette, Indiana 47907
S.R. STERNBERG (79), Machine Vision International, Ann Arbor, Michigan 48104
S.L. TANIMOTO (31), Department of Computer Science FR-35, University of Washington, Seattle, Washington 98195
D.L. TUOMENOKSA[3] (133), School of Electrical Engineering, Purdue University, West Lafayette, Indiana 47907
L. UHR (19), Computer Sciences Department, University of Wisconsin-Madison, Madison, Wisconsin 53706

[3]Present address: AT&T Information Systems, Holmdel, New Jersey 07733.

Preface

New technologies modify the connections between technical and cultural phenomena in our society because the introduction of new techniques in the production and consumption of goods and in the daily life strongly influence the nature of interhuman relationships. An interesting example was the war unit of man, horse, and lance made possible after the introduction of the stirrup (which guaranteed stability of the horseman); this unit has certainly modified the management of land and vassals by means of the new available military strategy.

In our age the computer has changed many traditional ways of thinking—it has even halted thinking altogether in some cases—and it is increasingly recognized that the possibilities offered by the very large scale integration of circuits and components have a major impact on the way in which the next generations of computing systems will be designed and built.

Within the field of computer image analysis we had decided to explore and exchange some of the existing ideas for building new multicomputer systems at Polignano, south of Bari, during June 1983. This workshop was fifth of a series started in 1979 (Windsor), 1980 (Ischia), 1981 (Madison), and 1982 (Abingdon); another workshop was held at Tucson in 1984.

This book contains all the material discussed at Polignano and some other related work where the influence of integrated microelectronics, in two and three dimensions, on the development of sophisticated computer architectures was particularly highlighted. As in past workshops, the multidisciplinary nature of pictorial information processing has been preserved: relationships between algorithms and architectures of multicomputer structures are constantly underlined in almost every chapter of the book.

The first chapter provides a general approach to the modelling of spatial and temporal aspects of concurrent systems, the next chapter discusses new sophisticated configurations of processors mapping optimal graphs, and the following chapter examines the role of image buffers for filling the gap between a cellular array and the many, possible, sequential host computers. The fourth article debates a possible MIMD framework for high speed computation underlining the interconnection conflicts. Next, an overall analysis, by simulation, of the different general image tasks follows where both SIMD and MIMD models are considered, and then the mathematical morphology approach is revisited and the most recent architecture for implementing it is presented. Hierarchical computation by pyramid architectures is first used for extracting corners and then described in a recent project; a variable configuration system (PASM) is discussed at work on contour extraction, and later some application projects using the CLIP4 system are critically analyzed. Finally, innovative 3-D microelectronics is presented for the construction of an image processor and superlattices are considered as possible shift registers, and a new project (CLIP7) is introduced. The last chapter deals with the problems encountered in the design of image processors to be carried on satellites.

Simulations of expected performances, evaluation of existing and proposed systems, and suggestions for new, different, multi-processor configurations are the main lines along which the general discussion on nonconventional image processing systems proceeds aiming to achieve the highest possible throughput only limited by today's technology and by the existing algorithms for performing both low and high level vision tasks.

I hope that this series of workshops may continue as a means to improve and deepen human communication which is absolutely necessary to prevent computers from executing tasks in a manner which may become too remote from the real needs of man.

Chapter One

*Specification Languages for Modelling Concurrency**

Pierpaolo Degano and Ugo Montanari

1. INTRODUCTION

Recent technological advances make it possible to integrate several hundreds of thousands of active elements in a single chip. Thus, the most critical stage of the chip production process, in terms of both overall cost and required time, is now layout design rather than circuit fabrication. For instance, according to Southard [1], fabrication of about 1000 units of a moderately complex circuit is profitable, but it takes 10,000 units to make its design rewarding.

As a consequence, interest has grown in the study of methodologies, languages, and tools for making VLSI circuit design cheaper, faster, and more reliable [2–4]. The traditional process of layout design is hierarchically organized into several phases, usually the following:

(i) logic design,
(ii) electrical design,
(iii) geometrical design.

Each phase is then subdivided; e.g., the geometrical design may consist of planar stick, grid, and layout levels.

The design is refined from higher to lower phases and levels (top–down), the goal being the optimization of some measure of complexity, which in the end usually corresponds to area minimization. Thus, synthesis and optimization techniques are required. A bottom–up step involves analysis and is performed to check the consistency of a top–down step.

* Work partially supported by Consiglio Nazionale delle Ricerche, Progetto Finalizzato Informatica, Obiettivo Cnet.

INTEGRATED TECHNOLOGY FOR PARALLEL IMAGE PROCESSING
ISBN 0-12-444820-8

In practice, the electrical design phase is often an unnecessarily complex intermediate step, which can be skipped. But sometimes it is difficult to specify the intended behaviour at the logical level (e.g., when it involves timing constraints); then the electrical design is given the first consideration. Hence, phases (i) and (ii) may be alternative, and the traditional process of layout design may be depicted by a **Y**-shaped diagram [5].

Progress in the field of layout design clearly requires increased automation. Many approaches are possible.

(i) *Computer-aided design.* Increasingly sophisticated tools for analysis, simulation, and synthesis are provided by a CAD environment, but only the mechanical aspects of design are automatized. All the relevant creative decisions are made by the designer.

(ii) *Expert systems.* Some of the design rules are made explicit and stored in an expert system. Thus, several design decisions are suggested by the expert system; only occasionally are these questioned and changed by the designer.

(iii) *Silicon compilers.* The functional behaviour of the circuit is specified in a formal language, and a compiler directly translates the specification into a layout.

(iv) *Formal solution of special cases.* Layouts for interesting classes of algorithms can be formally represented and studied. Typical cases are matrix multiplication and fast Fourier transform. The VLSI realization of several algorithms proposed for image processing can potentially be approached in this way.

A constant feature of all the above approaches is the need for a formal specification language, capable of expressing all the desired aspects of circuit behaviour at the required level of abstraction. This is even more important for highly parallel circuits, in which the behaviour is intrinsically complicated and a succinct description is crucial. Moreover, the lack of a suitable specification language forces the designer to sneak relevant aspects of the behaviour into lower phases, making the whole design process confused and error prone.

Many specification languages for concurrent systems have been suggested for VLSI applications, e.g., Petri nets and Milner's (*Syncronous*) *Calculus for Communicating Processes.* A recent Ph.D. thesis by Cardelli [6] reviews the subject and proposes a connection structure among modules that works at all levels, thus supporting a natural top–down methodology. Cardelli describes the functional behaviour of the circuit by means of a model called *Clocked Transition Algebra.* A graph structure is used to represent the topological structure of the circuit at this phase. The concept of homomorphism is used to map this specification into an electrical description using the so-called *Connector-Switch-Attenuator* model of electronic circuits. Further applications of homomorphisms lead to planar sticks, grids, and finally layouts. Another language for circuit description is proposed in Milne [7].

In previous work, the present authors developed the formalism called *Grammars for Distributed Systems* (GDS), an operational model for concurrent distributed systems based on graphs and graph rewriting systems (see Castellani and Montanari [8] for a first description and Degano and Montanari [9] for a complete presentation, including a precise treatment of infinite and fair computations).

Our graphs depict both the spatial and the temporal aspects of the systems they specify. We use the notion of processes, of ports spatially connecting them, and of temporal dependency between two processes, so that a second process can start only when the first is terminated. A graph also contains the history of the computations performed so far, represented as a temporal and spatial structure of events.

As a specification language, GDS combine the suggestiveness of the pictorial representation, the adequacy of the modular structure of VLSI top–down design methodology, and the firm ground of a formal semantics.

In the following sections we introduce a sequence of relevant concepts that lead to a complete but informal description of GDS. Each step contains a description of a self-contained formalism that is capable of expressing the behaviour of a system at some level of abstraction.

2. LABELLED TRANSITION SYSTEMS

A simple model that expresses only the concepts of sequentiality and nondeterminism comes first. Given a set of states, a *labelled transition system* [10] associates a label that describes an *observable atomic action* to every possible transition. An observable atomic action is a basic, indivisible interaction of the system with the environment in which it operates. If the atomic action a transforms state S_1 into state S_2, we write

$$S_1 \overset{a}{\Rightarrow} S_2$$

We can now define a *big-arrow relation* $\boxed{}\!\!>$ as the reflexive transitive closure of $\Rightarrow$, labelled with strings of atomic actions. Formally, the big arrow is defined by two inference rules:

$$S \; \boxed{\lambda}\!\!> \; S$$

$$\frac{S_1 \; \boxed{w}\!\!> \; S_2 ; \; S_2 \overset{a}{\Longrightarrow} S_3}{S_1 \; \boxed{wa}\!\!> \; S_3}$$

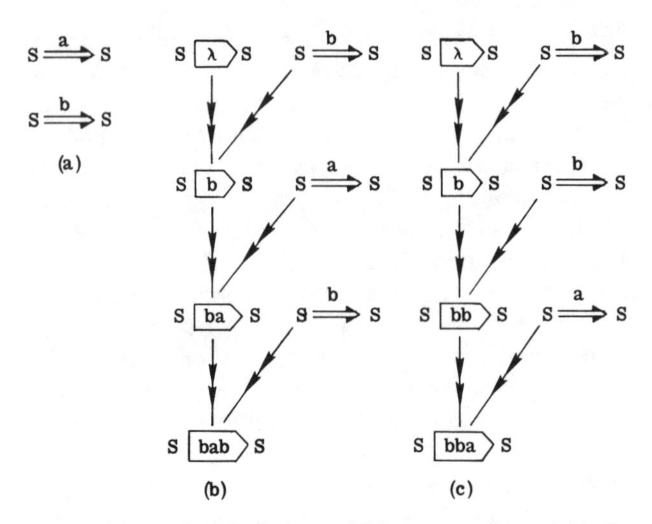

Fig. 1(a) The transitions of a labelled transition system. (b) and (c) Computations.

where λ is the empty string, and string wa is obtained by concatenating a string w with character a. The proof trees generated in this way are essentially the computations of the formalism. For example, if we have the transitions in Fig. 1(a), we can show that two computations labelled by *bab* and *bba* both transform S into S. In fact, we can construct the proof trees in Fig. 1(b) and (c), respectively, by using the above inference rules.

3. CONCURRENT TRANSITION SYSTEMS

In our next model which introduces concurrency, the state has an internal structure: It is a set of states of *concurrent processes* (or simply *processes*). In this model, an atomic action transforms a set of processes into a new set of processes. We call this the model of *concurrent transition system.*

In Fig. 2(a) we see two transition rules, in which a single arrow →, which represents the transition of a part of the system, replaces the double arrow ⇒. This is because we want to distinguish between the global state and the state of a set of process of the system. Transition rules, in general, involve only a part of the state, the global state being in a sense inaccessible.

How is the big arrow defined in this case, and in particular, what kind of label does it have? Notice that the two transition rules, involving disjoint sets of processes, may be concurrently applicable; thus a sequence of atomic actions is not expressive enough. We suggest a *partial ordering* of atomic actions.

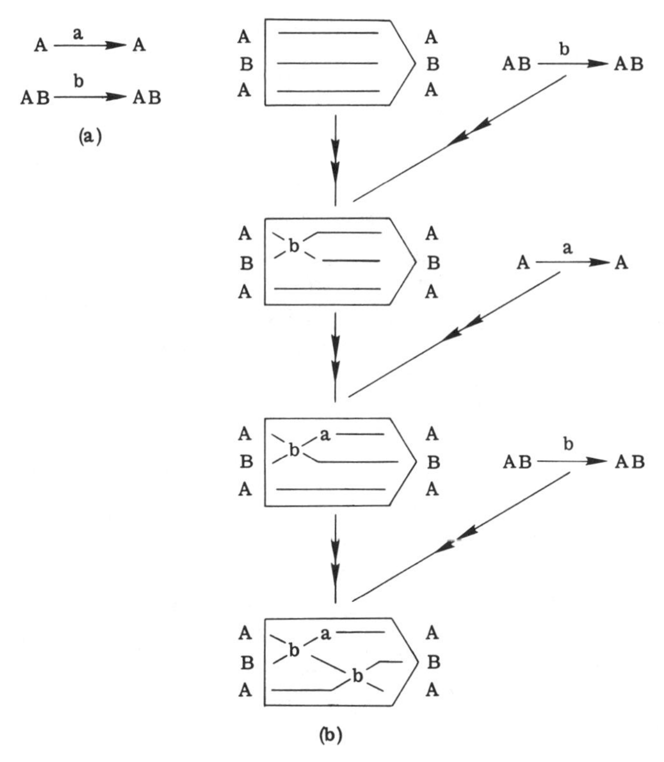

Fig. 2(a) The transition rules of a concurrent transition system. (b) A computation.

In Fig. 2(b) we see a computation in this model. The big arrow relates two states, i.e., two sets of processes. Its label is a partial ordering on the initial and final sets of processes and on the atomic actions performed in the computation. Notice that we can have more than one process with the same name in a state, and a transition rule can be applied to any of them.

Informally, the inference rules are as follows: A first rule with no antecedents relates two equal sets of processes with a *unity* partial ordering [see the first step in the computation in Fig. 2(b)]. Given two states (related by a big arrow) and a transition rule, the second inference rule makes it possible to replace, in the final state, the right-hand part of the transition rule for an occurrence of its left-hand part. At the same time, the partial ordering labelling the big arrow is augmented with the atomic action of the transition rule. Our example represents a system in which the A processes may evolve concurrently and may synchronize, one at a time, with the B process. Notice that the same last element in the computation in Fig. 2(b) can also be obtained by exchanging the last two steps. In fact, this element can be considered to represent, at a higher level of abstraction, both computations. The same result could not be obtained by using total orderings (i.e., strings) as labels, as shown by Fig. 1(b) and (c), in which two analogous computations generate different results (interpret S as ABA).

4. SYNCHRONIZED CONCURRENT TRANSITION SYSTEMS

The next step discusses the concept of synchronization. So far, a transition involving many processes is understood as a single, indivisible move. On the other hand, it can often be seen as the *composition*, through a *synchronization mechanism*, of a separate move for each process involved. Each possible move is described by a labelled production. The left-hand member of a production contains a single process, and its label is called a *communication protocol*. The synchronization mechanism is represented by a commutative partial function σ, mapping, if defined, any multiset of communication protocols into an observable action. The function σ, called the *synchronization function*, makes it possible to specify which productions can be composed in a single transition rule. This model is said to contain *synchronized concurrent transition systems*.

Figure 3 shows this version of our running example. We need to add to the previous rules a third inference rule. Given a multiset of productions, whose labels are mapped by σ in an observable action, the rule derives a transition rule labelled by this action. Its left member contains all the left members of the productions, and its right member is the (multiset) union of the right members.

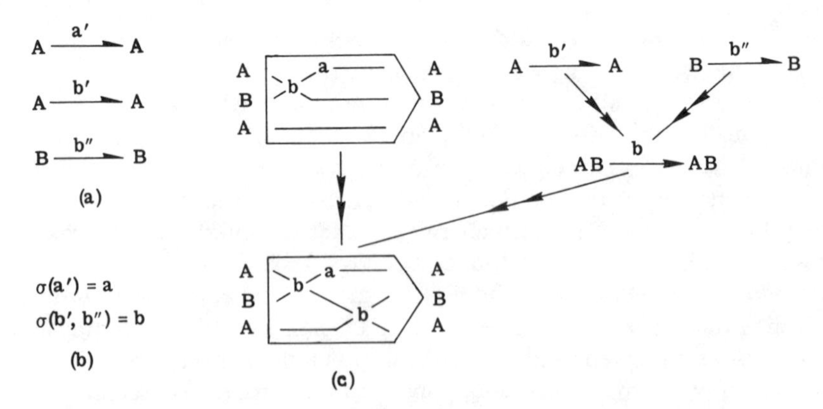

Fig. 3(a) The productions of a synchronized concurrent transition system (notice the half arrow). (b) The synchronization function. (c) A step in a computation.

The approach used here, setting out the evolution of a synchronized concurrent transition system in three steps—stand-alone (productions, half arrow), synchronization (transition rules, arrow), and global phases (computation step, big arrow)—is borrowed from the methodology developed by Astesiano [11], with the exception that in the third step he allows access to the global state.

5. DISTRIBUTED PORT-SYNCHRONIZED TRANSITION SYSTEMS

The above discussion does not consider aspects of distribution in space. In fact, even if a multiplicity of processes is assumed in a subsystem, the synchronization mechanism is essentially centralized. We now give a spatial structure to our systems by introducing the notion of adjacency between

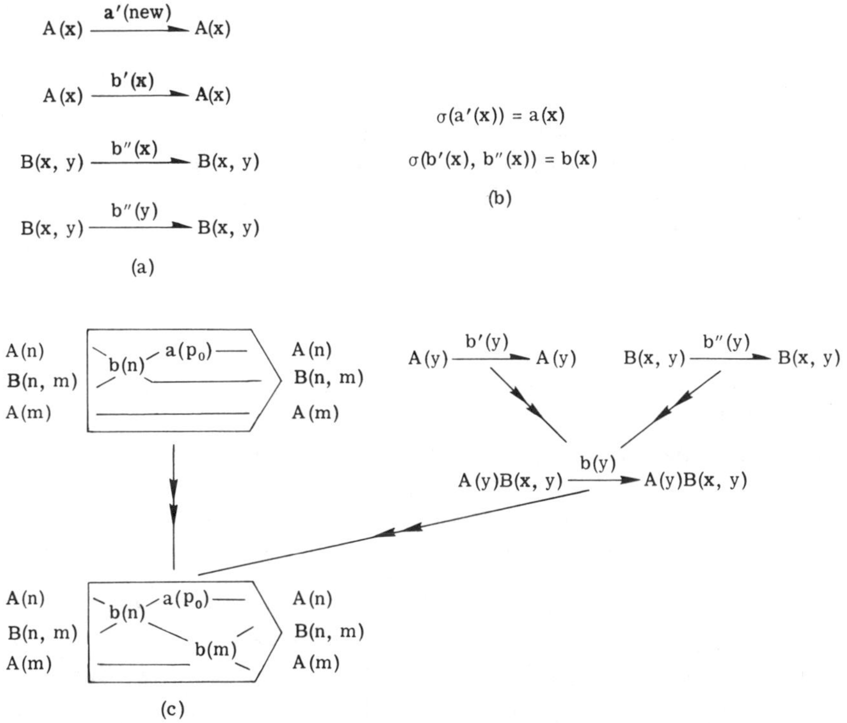
A(x) a′(new) A(x)
A(x) b′(x) A(x)
B(x, y) b″(x) B(x, y)
B(x, y) b″(y) B(x, y)
(a)
σ(a′(x)) = a(x)
σ(b′(x), b″(x)) = b(x)
(b)
A(n)
B(n, m)
A(m)
b(n)
a(p0)
A(y) b′(y) A(y)
B(x, y) b″(y) B(x, y)
A(y)B(x, y) b(y) A(y)B(x, y)
b(m)
(c)

Fig. 4(a) The productions of a distributed port-synchronized transition system. (b) The synchronization function (the same for all ports). (c) A step in a computation.

Identifiers x and y represent port variables, whereas m, n, and p_0 represent ports. Primitive *new* dynamically creates a new port when called.

processes. Two processes are *adjacent* if they are both connected to the same channel, or *port*, through which they can communicate. Only adjacent processes can be synchronized; thus ports are independent loci of synchronization. The synchronization mechanism is the same for all ports and is the same as in the previous model at the centralized level. These systems will be called *distributed port-synchronized transition systems*.

In Fig. 4 our example is enriched by a spatial structure. The state of our three-process system is represented by A(*n*)B(*n*,*m*)A(*m*); that is, we assume that the B process is connected to two distinct ports, *n* and *m*, where the two A processes are also connected. We assume the alphabet of the process names is ranked as follows: The symbols, 0_0, 0_1 . . . denote processes connected to no ports; A_0, A_1, . . . to one port; and so on. Communication protocols and actions must take place on a port. New ports can be dynamically created by a production, possibly together with new processes connected to them. In our example, the first production in Fig. 4(a) creates a new port through the primitive *new*. By the same token, a port can become disconnected from any process after a computation step and should in this case be disregarded.

Notice that in productions, synchronization functions, and transition rules we use port variables, which are instantiated to actual ports in performing a step of computation. This amounts to an assumption of homogeneity for space in our model.

6. DISTRIBUTED SYNCHRONIZED TRANSITION SYSTEMS

So far, only processes connected to the same port can be synchronized. This is a limitation. We can easily give a richer, more dynamic structure to the synchronous parts of an asynchronous system by allowing the synchronization of many ports through productions that are labelled by two (or more) protocols [see, for instance, the last production in Fig. 5(a)]. These protocols must be exchanged simultaneously; that is, all the processes involved in the synchronization on ports must appear in the left-hand part of the corresponding transition rule. See, for instance, the transition rules in Fig. 5(b). We call these systems *distributed synchronized transition systems*.

This new possibility allows us to make the synchronization mechanism on ports much simpler than the one described in the previous model, because we can always model an arbitrary synchronization mechanism through a suitable process (*synchronizer*). Thus we assume that no more than two processes at a time are connected to the same port. Furthermore, for the sake of simplicity, we assume that every synchronization involves exactly two identical protocols and thus we omit the specification of σ considering it as implicitly defined.

$$A(x) \xrightarrow{\tau} A(x)$$

$$A(x) \xrightarrow{b(x)} A(x)$$

$$B(x, y) \xrightarrow{b(x)} B(x, y)$$

$$B(x, y) \xrightarrow{b(y)} B(x, y)$$

$$B(x, y) \xrightarrow{b(x)b(y)} B(x, y)$$

(a)

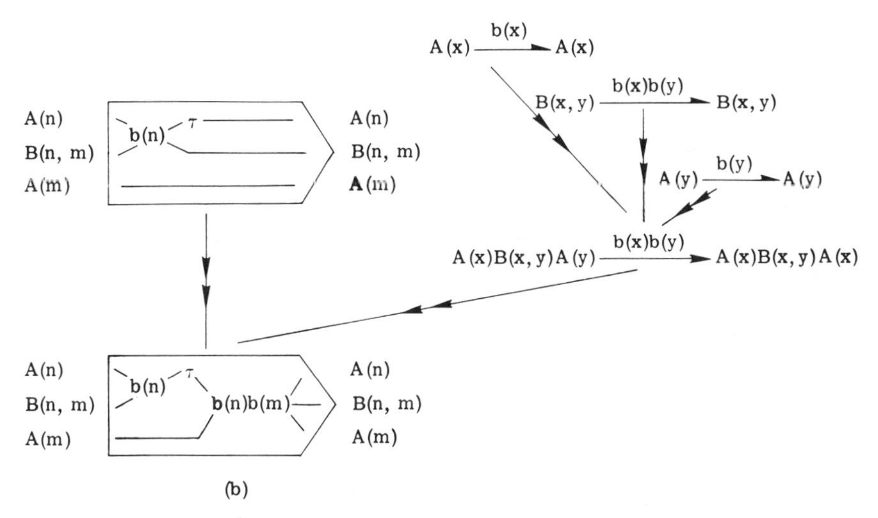

Fig. 5(a) The productions of a distributed synchronized transition system. (b) A step in a computation.

When two or more ports are synchronized, the labels of productions and transition rules and the elements in the partial order describing a computation (big arrow) can contain two or more simultaneous actions. Thus, the labels are better defined as tuples of (simultaneous) actions, or *events*. The event that contains no action is called τ; it refers to the absence of ports [see, for instance, the first production in Fig. 5(a)].

To exemplify how a synchronizer works see Fig. 6, the case of four processes connected through the same port. The A_1 and A_2 processes behave as senders (protocol p), and the A_3 and A_4 processes behave as receivers (protocol $\bar{p}$). Function σ synchronizes only a sender–receiver pair. In Fig. 7, sender–receiver synchronizations are guaranteed by synchronizer D. Notice

$$A_1(x) \xrightarrow{p(x)} A_1'(x)$$

$$A_2(x) \xrightarrow{p(x)} A_2'(x)$$

$$A_3(x) \xrightarrow{\bar{p}(x)} A_3'(x)$$

$$A_4(x) \xrightarrow{\bar{p}(x)} A_4'(x)$$

(a)

$$\sigma(p(x), \bar{p}(x)) = a(x)$$

(b)

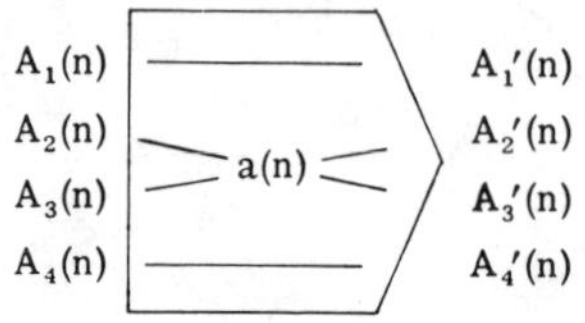

$A_1(n)$ $A_2(n)$ $A_3(n)$ $A_4(n)$ — $a(n)$ — $A_1'(n)$ $A_2'(n)$ $A_3'(n)$ $A_4'(n)$

(c)

Fig. 6(a) The productions of a distributed port-synchronized transition system. (b) An element of a computation.

$$A_1(x) \xrightarrow{p(x)} A_1'(x) \qquad D(x, y, z, w) \xrightarrow{p(x)\bar{p}(z)} D(x, y, z, w)$$

$$A_2(x) \xrightarrow{p(x)} A_2'(x) \qquad D(x, y, z, w) \xrightarrow{p(x)\bar{p}(w)} D(x, y, z, w)$$

$$A_3(x) \xrightarrow{\bar{p}(x)} A_3'(x) \qquad D(x, y, z, w) \xrightarrow{p(y)\bar{p}(z)} D(x, y, z, w)$$

$$A_4(x) \xrightarrow{\bar{p}(x)} A_4'(x) \qquad D(x, y, z, w) \xrightarrow{p(y)\bar{p}(w)} D(x, y, z, w)$$

(a)

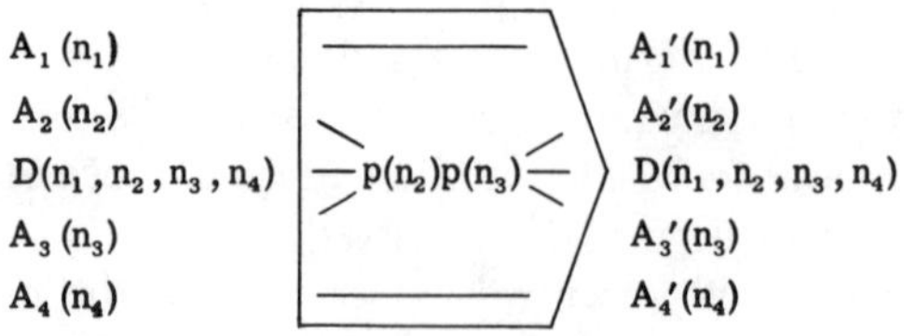

$A_1(n_1)$ $A_2(n_2)$ $D(n_1, n_2, n_3, n_4)$ $A_3(n_3)$ $A_4(n_4)$ — $p(n_2)p(n_3)$ — $A_1'(n_1)$ $A_2'(n_2)$ $D(n_1, n_2, n_3, n_4)$ $A_3'(n_3)$ $A_4'(n_4)$

(b)

Fig. 7(a) The productions of a distributed synchronized transition system. The D process acts as a sender–receiver synchronizer. (b) An element of a computation.

that in Fig. 6(c) the action is $\sigma\ (p(n), \bar{p}(n)) = a(n)$, whereas in Fig. 7(b) the event consists of action p on port n_2 and action $\bar{p}$ on port n_3.

It is not difficult to write productions for sender–receiver synchronizers with any number of ports. It is also possible to imagine productions that define a large variety of synchronizers.

7. TAIL–RECURSIVE GRAMMARS FOR DISTRIBUTED SYSTEMS

We present now the first version of our graph model, which we call *tail-recursive Grammars for Distributed Systems.* We want a more traditional derivation structure, in which, as inTuring machines or in phrase structure grammars, there is only one notion of unlabelled computation or derivation. Recall that in a distributed synchronized transition system, a computation is a sequence of pairs of (initial and final) sets of processes related by a partial ordering of events. The initial set is always the same and thus can be factored out and treated as the *initial graph* of the grammar. Therefore, we can see a computation as a sequence of partial orderings that relate *subsystems*, i.e., events and processes. A partial ordering is intended to express the *causal* relation between two subsystems; i.e., all the events that are smaller than a subsystem are exactly those that have caused it. Pairs of subsystems that are not causally related are called *concurrent*. Finally, the ports and the connections between subsystems and ports are parts of the graph structure. In Fig. 8(b) we see a four-step computation.

We use the following conventions for our graphs. The vertical dimension of the page represents time, flowing from top to bottom, and the horizontal

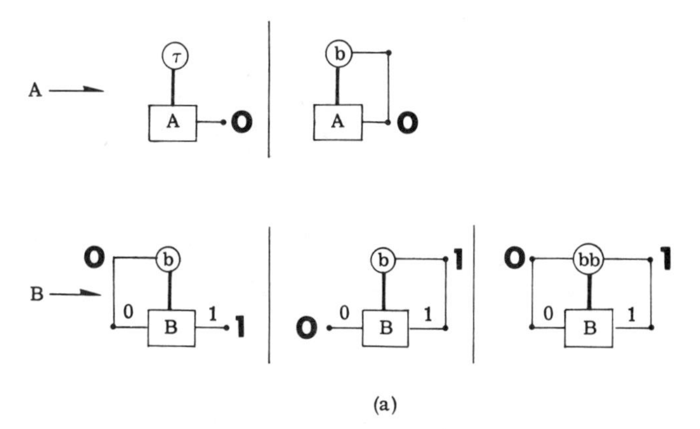

Fig. 8(a) The productions of a tail-recursive GDS corresponding to those in Fig. 5(a).

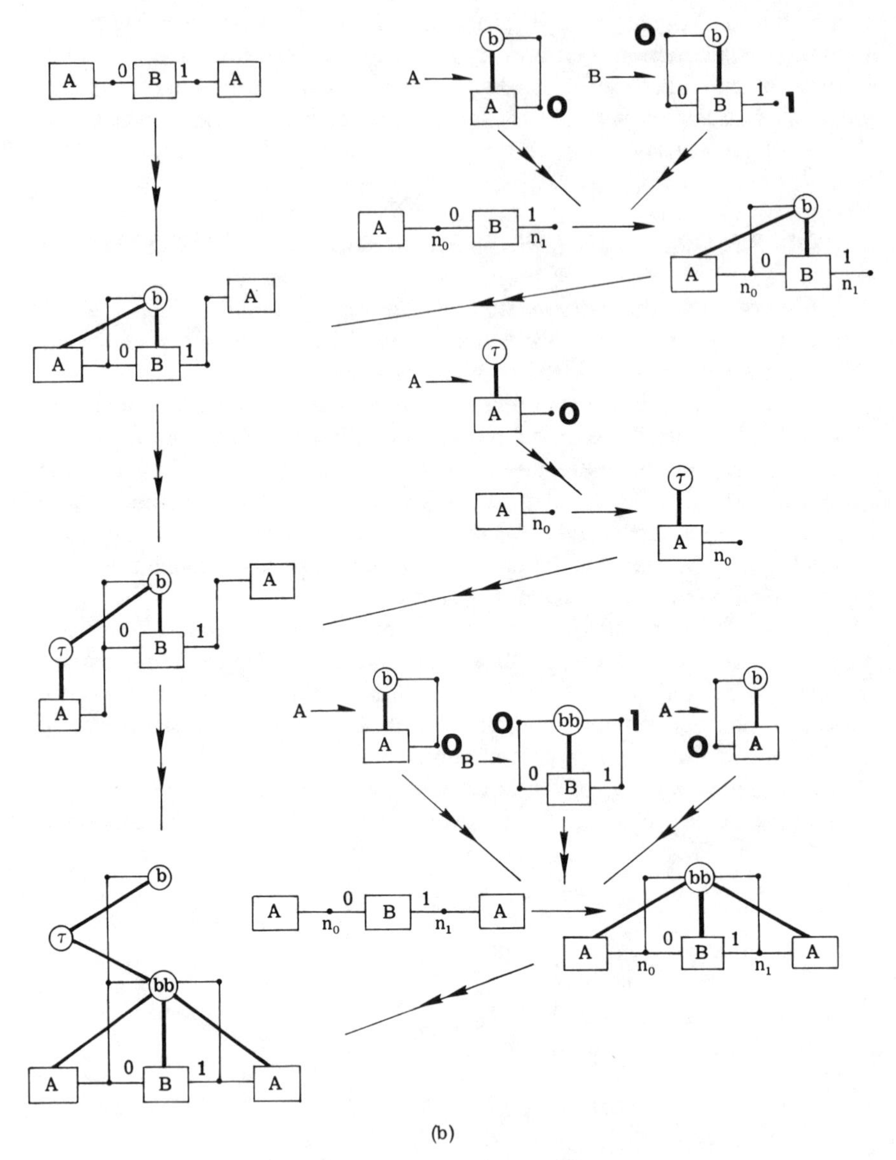

Fig. 8(b) A computation.

dimension presents space. Thus ports, which have only a temporal extension, are usually represented as vertical lines (of medium thickness). Events (resp., processes) are circles (resp., boxes); connections between subsystems and ports are thin horizontal lines; and intersections with ports are dots. Boxes contain process *types*, and circles contain the tuple of actions that form the event. Notice that, in general, the ports to which a subsystem is connected cannot be permuted. The connections of any box are numbered when necessary with small numerals. Since there is a correspondence between actions and the ports at which they happen, every action is drawn close to the connection of its corresponding port. Finally, the partial ordering is represented by its Hasse diagram, drawn from top to bottom with thick, nonhorizontal lines.

The productions of our previous model translate in a straightforward way to tail-recursive GDS productions [see Fig. 8(a)]. The left member of a production is simply a process type, and the right member is a graph with a single event. We also need to establish a correspondence between the ports of the left member and (some of) those of the right member. For this purpose, we label the relevant ports in the right member with large numerals and call them *external ports*.

Transition rules are analogously translated into *rewrite rules* [see the right part of Fig. 8(b)]. Corresponding ports in the left and right members are marked by the same name.

The inference rules that make it possible to derive rewriting rules from productions and to apply rules to graphs are quite naturally expressed in terms of graph rewriting concepts. In fact, given the left member of a rewriting rule and a production for every process in it, the right member is obtained in two steps, as follows. First, actual ports are matched against formal ports, and every process is replaced by its definition. Second, all the events of the productions are merged into a single event, checking that protocols on the same port match in pairs.

The application of rules to graphs is straightforward, because one has only to match the left member of the rewrite rule against a subgraph and to replace the right member for it.

8. GRAMMARS FOR DISTRIBUTED SYSTEMS

We now introduce full recursion in the productions that define our GDS. In the graphs, that we call *distributed systems*, the causal relation can hold between processes, too; i.e., a process is allowed to start only when there is no process preceding it. An example that takes advantage of full recursion is depicted in Fig. 9. It models the classical primitives of *fork* and *join*. The first production in Fig. 9(a) creates three processes, the first two of which are

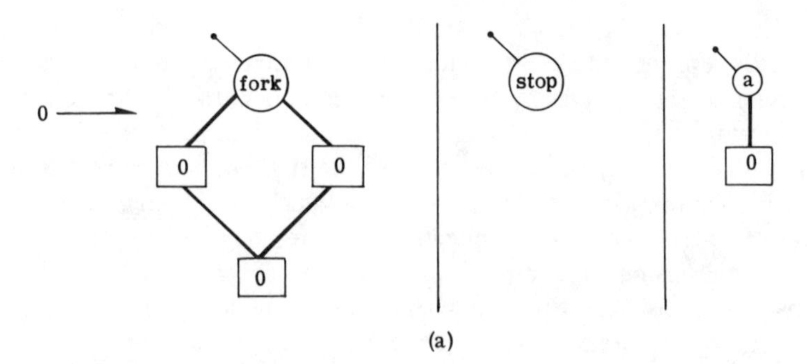

(a)

Fig. 9(a) The productions of a GDS modelling *fork* and *join* primitives.

immediately executable, and the third becomes ready only when the former processes are terminated by the second production.

Of course, in a distributed system, no process can precede an event in the partial ordering; and to perform a step of computation, a rewriting rule can be applied only to processes that are minimal in the partial ordering when events are removed.

9. TWO EXAMPLES

In Fig. 10 we see a more suggestive example, modelling a *cell*. A cell has two ports, an input port labelled 0 and an output port labelled 1. A cell has states modelled by the process types B(j), $j = 0, 1, \ldots$. The first parametric production describes the input operation; the second production models the output operation. The most natural way of joining two cells is to connect the output of one cell and the input of the other to the same port. In this case only one event can occur at a time on the connecting port, i.e., writing the contents of the first cell in the second. This is the situation depicted by the two central processes in the initial graph of the computation in Fig. 10(b). A second type of junction connects two outputs to the same port (the rightmost pair). In this case, synchronization can take place only if the two cells have the same content. Conversely, if two cell inputs are connected to the same port, any value can be nondeterministically loaded on both cells (the leftmost pair). The first step in the computation applies a rewriting rule derived from the second parametric production instantiated by $j = 1$ and the first production instantiated by $j = 0$ and $i = 1$. Notice that the rightmost process pair has no applicable rule in the initial graph.

A sequence of cells in which the output of one cell and the input of the next cell are connected to the same port, might seem to resemble a shift register.

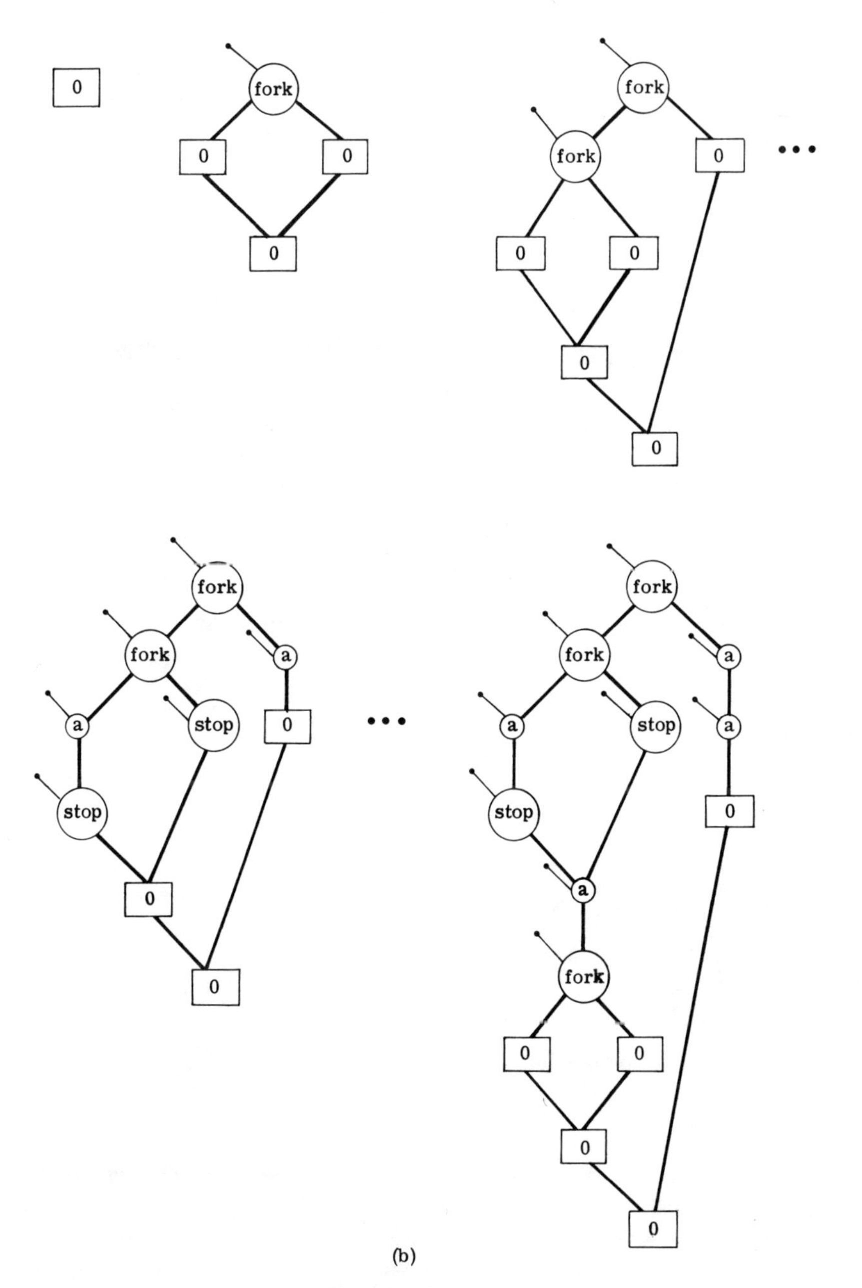

(b)

Fig. 9(b) A computation.

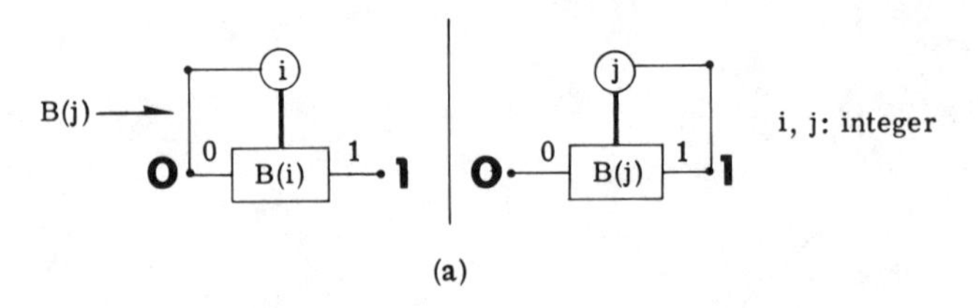

(a)

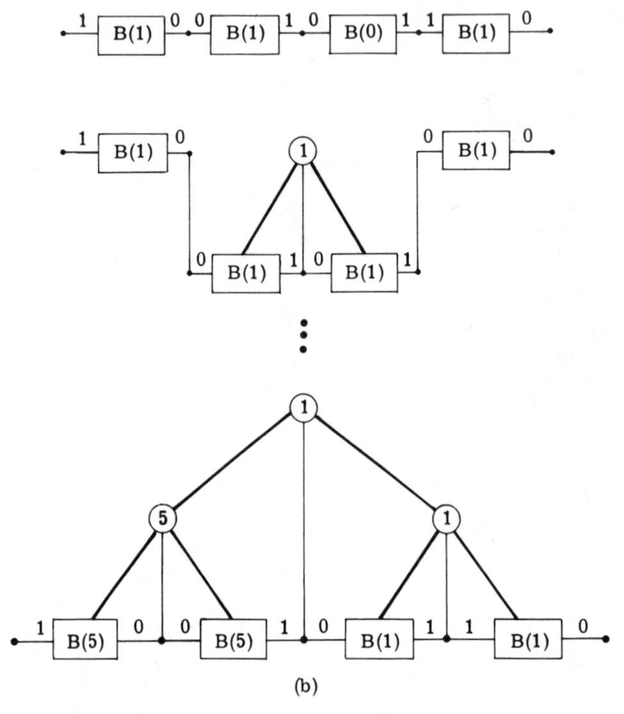

(b)

Fig. 10(a) The productions of a GDS modelling a cell. (b) A computation.

But this is not the case, because the productions of our grammar make it possible to propagate the content of the cells asynchronously; thus a cell can lose some value, and some other value can be loaded on more than one cell. However, a shift register is easily specified (see Fig. 11). The first production schema of Fig. 11(a) describes the evolution of a cell that, at the same time, reads on its left (0) port and writes on its right (1) port. Thus, the content of the whole register, regardless of its length, is shifted by the application of a

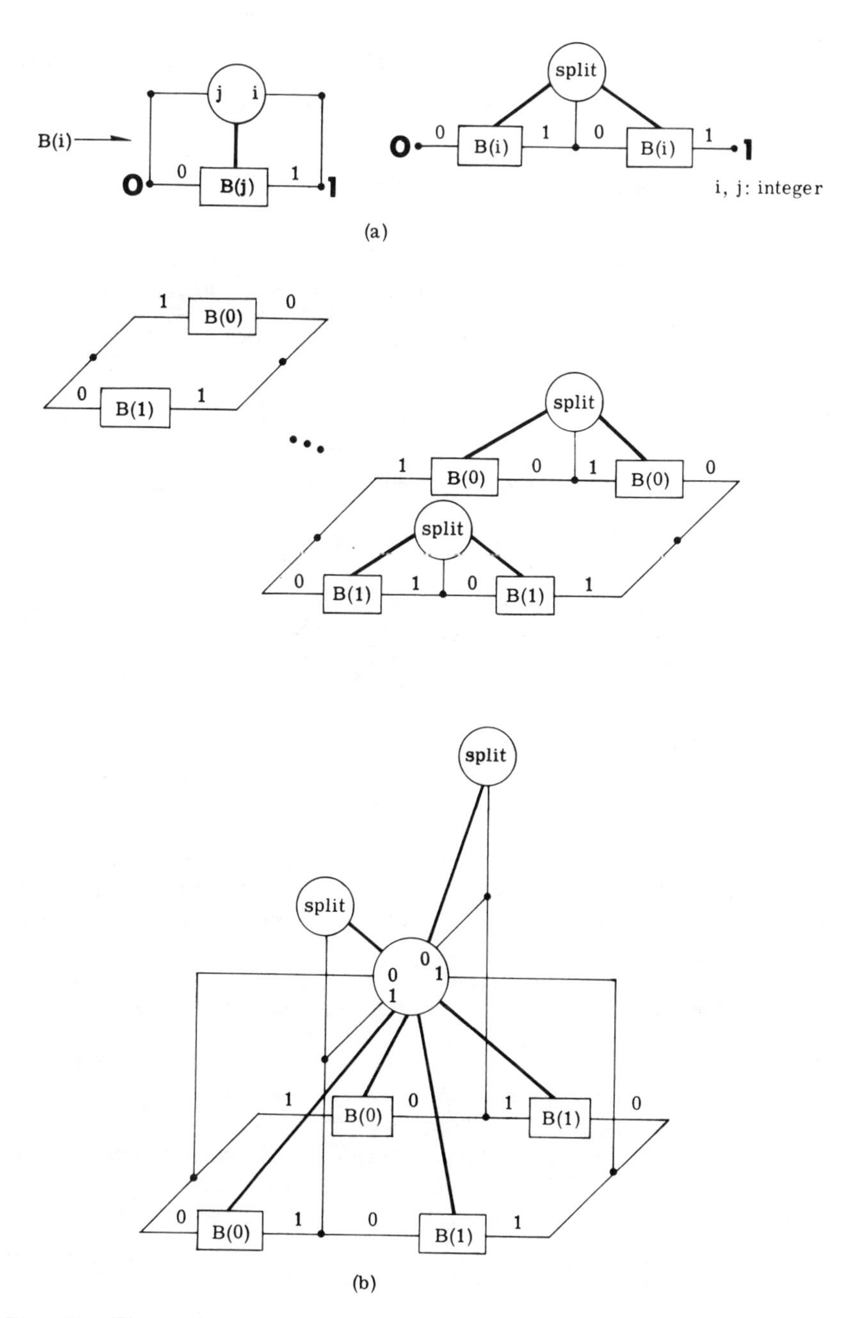

Fig. 11(a) The productions of a GDS modelling a shift register with self-duplicating cells. (b) A computation.

single rewriting rule. To have more fun, we added a second production schema that can asynchronously duplicate a cell. In Fig. 11(b) we see a computation of a circular (counterclockwise) shift register.

ACKNOWLEDGEMENTS

The authors wish to thank Ilaria Castellani for collaboration on early work on the subject, Sonia Raglianti for accurately typing the manuscript, and Nedo Iacopini for drawing the figures.

REFERENCES

[1] Southard, J. R. (1983). MacPitts: An approach to silicon compilation. In [3], pp. 74–82.
[2] Mead, C., and Conway, L. (1980). *Introduction to VLSI Systems*, Addison-Wesley, Reading, Massachusetts.
[3] *Computer*. (1983). Special issue on new VLSI tools, **16**, 12.
[4] Commission of the European Communities. (1983). ESPRIT 1983 workplan, sub-programme advanced microelectronics.
[5] Gojski, D. D., and Kuhn, R. H. (1983). Guest editors' Introduction, New VLSI tools. In [3], pp. 11–17.
[6] Cardelli, L. (1982). *An Algebraic Approach to Hardware Description and Verification*. Ph.D. Thesis, CST-16-82, University of Edinburgh, Edinburgh.
[7] Milne, G. J. (1982). CIRCAL—*A Calculus for Circuit Description*, Internal report CSR-122-82, University of Edinburgh, Edinburgh.
[8] Castellani, I., and Montanari, U. (1983). Graph grammars for distributed systems. In *Proc. 2nd Int. Workshop on Graph-Grammars and their Applications to Computer Science* (H. Ehrig, M. A. Nagel, and G. Rozenberg, eds.). LNCS, **153**, Springer-Verlag, Berlin, pp. 20–38.
[9] Degano, P., and Montanari, U. (1983). *A Model for Distributed Systems Based on Graph Rewriting*, Note Cnet 111, Computer Science Dept., University of Pisa, Pisa.
[10] Plotkin, G.D. (1983). An operational semantics for CSP. In *Proc. IFIP TC 2-Working Conference: Formal Description of Programming Concepts II* (D. Bjørner, ed.), North-Holland, Amsterdam, pp. 199–223.
[11] Astesiano, E., and Zucca, E. (1981). Semantics of CSP via translation into CCS. In *Proc. MFCS '81* (J. Gruska and M. Chytil, eds.). LNCS, Springer-Verlag, Berlin, pp. 172–182.

Chapter Two

Augmenting Pyramids and Arrays by Embossing Them into Optimal Graphs to Build Multicomputer Networks

L. Uhr

1. INTRODUCTION

This paper explores how different types of serial computers, arrays, pyramids, networks, and other resources can be combined into relatively well-structured multi-computer systems. It attacks this general problem by concentrating on the particular problem of how pyramids and arrays of very large numbers of (typically relatively small) parallel computers can be combined with a much smaller number of (typically relatively much more powerful) serial computers into a closely point-to-point-linked multi-computer network. It focusses on such pyramid-net networks that are built from the optimally dense and symmetric Moore Graphs, in particular the (10, 3, 2) [nodes = 10, degree = 3, diameter = 2] Petersen Graph and the (50, 7, 2) Singleton Graph.

Two types of constructions are presented and examined:

(a) A single Moore Graph some of whose nodes are processors, while others are memories;

(b) A single Moore Graph some of whose nodes are pyramids or arrays, while the rest are traditional serial computers.

2. INTRODUCTION TO ARRAYS, PYRAMIDS, NETWORKS, AND OPTIMAL GRAPHS

Today's rapidly improving Very Large Scale Integration (VLSI) technologies will soon make feasible very large multi-computers with thousands, or even millions, of processors [1]. This paper focuses on two distinct types of

INTEGRATED TECHNOLOGY FOR
PARALLEL IMAGE PROCESSING
ISBN 0-12-444820

systems—synchronous, lock-step and asynchronous, independent—that have been proposed in recent years. The synchronous systems include several large parallel arrays, each with over 1000 computers which have been completed, and several pyramids of arrays that are now being built. Most asynchronous systems that have been built to date have fewer than 16 computers, but several have 50 to 80 computers. Systems have been partially designed or proposed that contain thousands, or even millions, of computers.

But virtually nothing has been done toward designing multi-computer networks that combine several different types of computer resources into a single system in which all the parts can work together, in parallel, with relative efficiency and speed. This paper begins to explore how such mixed networks might be configured.

Pyramids and arrays of large numbers of computers offer great promise for image processing, pattern perception, and other tasks. Arrays are especially appropriate when large amounts of information are to be processed using local operations that are applied everywhere, as in a television image. Pyramids are especially appropriate for executing programs that successively examine and transform this information into increasingly abstract representations of increasingly smaller amounts of successively more condensed, abstract, and symbolic information.

2.1 The Need to Augment Pyramids and Arrays with Networks

It appears that arrays and pyramids can benefit substantially from being combined with a Multiple Instruction-Multiple Data (MIMD) stream network of more powerful traditional serial computers (see Flynn [2] and Uhr [3], [4] for examinations of these issues). For example, for the perception of objects in motion, a pyramid can be used very effectively to process the continuing streams of images that are input into its base array. But pyramids appear to most people to be restrictive and limited at the higher levels, although this may be because appropriate parallel algorithms have not yet been worked out.

Rather than interrupt the bottom–up flow of processes (as successive images are streamed into and up the pyramid) to execute the top–down processes that will also be needed (e.g., to gather more information about particular regions of interest, to adjust thresholds and weights used at low levels, and to broadcast information), top–down processing can be accomplished by using the additional nodes of the larger network with which the pyramid has been combined. And at the so-called higher (more abstract and symbolic) levels of pattern perception, it is very difficult to use the local synchronous Single Instruction–Multiple Data (SIMD) stream (that is, with all computers being driven by a single controller and synchronized to execute the same instruction, albeit on different sets of data) capabilities of today's arrays and pyramids to process regions, symbols, and models of objects, by executing processes that today are typically executed on a traditional computer.

2.2 Very Large Arrays of Very Simple One-bit Computers

Three large arrays, each with substantially more than 1000 computers, have been built during the past few years. These include the 64 × 64 ICL DAP [5]; the 96 × 96 CLIP4 at University College, London [6]; and the 128 × 128 MPP, by Goodyear-Aerospace for NASA-Goddard [7]. Each of the 4000 to 16,000 computers in these multicomputers is directly linked to its four (DAP, MPP) or eight (CLIP) nearest neighbors, forming a two-dimensional grid. These are all SIMD systems; they are made up of very small 1-bit computers, each with only 32 bits (CLIP4) to 4096 bits (DAP) of memory.

But each of their thousands of computers is a true general-purpose computer that is capable of executing any program that can be written for any other computer.

These systems gain three or four orders of magnitude in speed and power because of the many thousands of computers used, at the relatively minor cost of slowdowns when they must be used bit-serially to process pieces of information that are larger than 1 bit (e.g., 4-bit weights, 8-bit grey-scale images or symbols, or 32-bit numbers).

Potentially even more important, these parallel arrays can be built as large as one desires, simply by adding more computers. The fabrication and linking of these simple computers is ideally suited to Very Scale Integration (VLSI) technologies. Therefore it will be feasible to build 512 × 512 and even 1024 × 1024 arrays during the next five to ten years.

2.3 Pyramids of Successively Smaller Layers of Arrays

Several papers have been written describing, examining the advantages of, and giving preliminary designs for pyramid multi-computers (e.g., Dyer [8], Tanimoto [9], Uhr [3], [10]). No pyramids have yet been completed, but one (designed by S. Tanimoto, University of Washington, and being built by Boeing Aerospace) is under construction and nearing completion. [In addition, V. Cantoni, University of Pavia, and S. Levialdi, University of Rome, are building a smaller system in Italy, and D. Schaefer is building a 16 × 16 based pyramid, using MPP chips, at George Mason University.]

The University of Washington pyramid will have a 64 × 64 array of 1-bit computers at its base, linked to 32 × 32, 16 × 16, 8 × 8, 4 × 4, 2 × 2, and 1 × 1 arrays. Each computer (with the exception of the computers in the base array) is linked to the four children directly below it in the next-larger array and to the one parent directly above it in the next-smaller layer. This pyramid has a single controller, but each array layer can be broadcast a different instruction to execute.

Note that a pyramid can be viewed as a stack of successively smaller arrays through which information can be sent, processed, transformed, and pushed ever closer together.

This kind of pyramid can be used as a two-dimensional pipeline to process, feature-detect, transform, compound, abstract, and recognize sequences of images streamed into its base array (e.g., by a TV camera). It can also be used to pass messages between distant nodes, up and then back down through the pyramid. This reduces the $O(N)$ diameter on an $N \times N$ array to $O(\log N)$, the much, if not crucially, better diameter of the pyramid.

For example, a 1024 × 1024 4-connected array needs 2047 shift operations to move the most distant pieces of information together. In sharp contrast, a 1024 × 1024-based pyramid needs only 19 (if it shifts one of these pieces up to the apex node and then back down to the node that contains the other piece of information). This already acceptably small number can be cut further, to only 9 shift operations, simply by shifting both pieces of information up to the apex node.

2.4 Networks of Traditional Computers

A number of MIMD networks have been designed and constructed in which each computer is substantially more powerful and has its own controller (and thus each computer asynchronously executes different instructions) [1]. A large number of multi-computer networks have been designed (e.g., partially re-configurable networks such as Siegel's PASM [11]; Briggs *et al.*'s PM 4 [12]; and Mago's [13] and Despain and Patterson's [14] trees). A few of these (with at most 50 or 80 processors) have been built (e.g., Swan *et al.*'s CM★ [15]; Wittie's MICRONET [16]; Manara and Stringa's EMMA [17]. These networks typically use off-the-shelf 8-bit, 16-bit, or 32-bit computers, e.g., the Z80 (Zmob, Rieger *et al.* [18]), PDP-11 (Cm★), M68000 (PASM), or VAX-750 (Crystal, Cook *et al.* [19]).

The two largest networks actually built (Cm★), Swan [15] and the Genoa machine [17] have 50-80 computers. Most MIMD networks to date are prototypes with 3–16 computers (e.g., MICRONET, Wittie [16], Lipovski and Tripathi [20]). Thus, the MIMD networks contain 10^1–10^2 computers; the SIMD arrays (and the pyramids under construction) contain 10^3–10^4. Wittie [16] originally talked of building networks with 10^6 computers.

Siegel's long-term goal is 10^3. Researchers at Columbia University and at the Massachusett's Institute of Technology are also beginning to talk about 10^6. But it may be that a system in which each controller is shared by many very simple computers, all linked together in a simple near-neighbor fashion, can contain 10^2-10^3 more computers for the same cost [21].

2.5 Optimally Dense Moore Graphs and Good Network Characteristics

The transfer of information between computers (often called *message-passing*, whether it be of parameters, data, programs, or other sets of information) turns out to be, at least with today's implementations, the most time consuming function of a network; and most network topologies have been chosen to minimize message-passing problems. Usually suggested as the most important criteria for choosing a particular network topology are the packing of as many nodes as possible as close together as possible, and the ease at finding short, addressable paths between pairs or sets of nodes that must exchange information. The density of the network (i.e., the number of nodes in a graph of given diameter and degree, where diameter is the longest shortest distance in the graph, and degree is the maximum number of edges at a single node) is usually taken to be the best measure of this characteristic.

It has been proved that the Moore Graphs (see Hoffman and Singleton [22]) are the only graphs that achieve optimal density. The Moore Graphs include the trivial (for our purposes) complete graphs, (in which every node is linked directly to all other nodes, giving too many links and ports to be realized in actual computers), polygon graphs (in which nodes are linked in a ring, giving graphs with diameters that are far too long) and only three other graphs: the (10, 3, 2) Petersen Graph, the (50, 7, 2) Singleton Graph, and the possible, but not yet discovered (3250, 57, 2) graph.

This paper concentrates on the Petersen and Singleton Graphs as bases for multicomputer networks that combine SIMD (pyramid or/and array) and MIMD (traditional Von Neumann serial computer) components. But the kinds of design considerations used with these graphs might be used with whatever topology one chooses.

3. COMBINING SIMD PYRAMIDS (OR/AND ARRAYS) WITH MIMD NETWORKS INTO OPTIMAL MOORE GRAPHS

The 10-node (10, 3, 2) Petersen Graph and the (50, 7, 2) Singleton Graph are especially attractive candidates for the basic network structure into which to combine SIMD (pyramid and array) and MIMD (von Neumann serial computer) components. (Any other graph topology might be chosen instead, but the rich variety of possibilities for just these two graphs is already too great to be explored here.

Each of these graphs can be used to form a variety of attractive SIMD-MIMD multicomputer networks. Some of these possibilities will be examined now.

The 10-node Petersen Graph is an attractive structure with which to start, because of its small size, relative simplicity, optimal density, symmetry, and other properties that are related to good network behavior (see Bondy and Murty [23] Uhr [1]).

4. BUILDING MULTIPROCESSOR SYSTEMS USING PETERSEN GRAPHS

The first example of a multiprocessor system is something of a diversion, but it is presented because of its simplicity. It breaks the traditional serial computer into its basic parts and uses the Petersen graph to link them back together. Rather than link several processors and several memories together over a single common bus, it is attractive to consider linking them together with a 10-node Petersen graph.

One elegant construction links six processors: Three large memories, each shared by two of the processors, and one very large disk, directly linked to the three memories. This can be effected by designating any node in the Petersen graph as the disk.

Then the three nodes directly linked to this disk node are the memory nodes, and the other six nodes are the processors. Each processor links to two other processors and to one memory; each memory links to two processors.

5. COMBINING ONE PYRAMID WITH A NETWORK OF NINE SERIAL COMPUTERS

The first pyramid-net construction simply puts a conventional MIMD computer at nine of the ten nodes of (10, 3, 2) and one pyramid or array at the tenth node. (Since the Moore Graphs are completely symmetric, it makes no difference which node is chosen for the pyramid.)

The pyramid is now directly linked to the three MIMD nodes that are separated from it by one node and through these nodes to the rest of the network (i.e., to the other six nodes, which must be reached via these three near neighbors and hence are two nodes away from them).

Each of the three links from the pyramid can come from a different subset of the pyramid's computers.

For simplicity in this first example, let us assume that each conventional MIMD computer is a 32-bit computer, and each SIMD computer in the pyramid is a one-bit computer. Therefore, three sets of 32 SIMD computers can be linked directly to the three adjacent MIMD computers. This might be carried out in the following way (note that this is but one of many possibilities).

Link the 8×8 array in the pyramid (the pyramid can have a base of any size, as we shall see) to two of the three MIMD nodes, as follows: Configure the 32 1-bit lines from each MIMD node into a 4×8 array, and pair the two to form an 8×8 array. Link each node in this 8×8 array to a 1-bit computer node in the 8×8 array of the pyramid.

Each of these pyramid computers is the 1×1 apex layer of the subpyramid that links below it. Thus, if the grand pyramid has a 128×128 base, each subpyramid has a 1×1, 2×2, 4×4, 8×8, and 16×16 layer below it.

The third MIMD node can be linked to whatever set of 1-bit pyramid computers seems most desirable, e.g., the 21 computers in the 4×4, 2×2, and 1×1 layers, plus 11 other judiciously chosen computers spread out in the lowest layer or layers. Or, alternatively, its lines can be linked to input and to output devices.

A large number of variations on these combining mechanisms are possible.

6. COMBINING $N(> 1)$ PYRAMID NODES WITH $M(= 10 - N)$ NETWORK NODES

There is no need to assign all but one node to the MIMD net. The ten nodes of (10, 3, 2) can be distributed in any way deemed desirable to pyramid and net nodes. The number of network nodes M can be multiplied by the number of bits B in each MIMD processor. This gives the number of 1-bit lines that can be used to link to the 1-bit SIMD nodes in the pyramid or array. Similarly, each SIMD computer is composed of N 1-bit computer nodes *in toto*. Usually, N will be much greater than MB; therefore, a sub-set of MB SIMD processors must be chosen to be the recipients of direct links.

It seems most attractive to spread these throughout the SIMD structure. But often in a pyramid there is a need to link nodes in high-level layers (e.g., layers 8, 9, 10, . . . in a large pyramid) with nodes in low-level layers (e.g., layers 2, 3, and 4). Alternatively, these extra links can go to input, output, and/or mass-memory devices. In an array it may be advantageous to link otherwise distant nodes together, from opposite sides. The architect has a great deal of freedom in placing these links, according to whatever combination of criteria is deemed most desirable.

The following are several specific SIMD–MIMD networks that seem especially desirable and worth considering seriously for possible construction:

(A) Combine five 64-bit MIMD nodes with a pyramid, as follows:
Take the 16 × 16 layer of the pyramid, and link the 64 computers in each of its 8 × 8 quadrants to a different MIMD node.
Link the fifth MIMD node to the 8 × 8 layer immediately above.

Now the lower, larger layers of the pyramid are linked to the MIMD nodes through the 256 subpyramids whose apexes are at the 16 × 16 layer; and the higher, smaller layers are linked through the 8 × 8 layer.

(A′) Combine five 64-bit MIMD nodes with a pyramid, as in (A) except that the 64 links to the fifth MIMD node from the higher layers of the pyramid are scattered differently; for example,

(1) Sixteen links go to the 8 × 8 array (e.g., to nodes)

```
 1, 1    1, 3   ...    1, 15
 3, 2    3, 4   ...    3, 16
                 .
                 .
                 .
15, 2   15, 4   ...   15, 16
```

(2) Sixteen links go to the 4 × 4 array (to all nodes, but permuted.
(3) Four links go to the 2 × 2 array.
(4) One link goes to the 1 × 1 array.
(5) Thirteen links go to scattered nodes in two large arrays next to the base of the pyramid (possibly to the center nodes of a 3 × 3 of subarrays in layer 2 and to the four center nodes of a 2 × 2 of subarrays in the next-higher layer 3, or in layer 4).

[Note that these kinds of scattered linkages serve a real purpose only if the MIMD computer is able to look at individual bits conveniently. For example, when an 8 × 8 subarray is loaded into the 64-bit computer, that computer should be able to treat this information as two-dimensional. When nine centers are linked to form a large array, it should be able to load or send information from any one, or several, of the centers. Therefore, the basic instruction sets of these computers must be extended beyond what is typical today (often, depending on the computer that is being used, this can be done relatively easily with microcode).]

(B) Combine one 64-bit MIMD node with a pyramid, as follows:
Use a pyramid that converges 3 × 3 from its 12 × 12 to its 4 × 4 layer, but 2 × 2 everywhere else.

Link every fourth node of the 12 × 12 layer (i.e., 36 nodes) to 36 lines of the MIMD nodes.

Link the other 28 lines from the MIMD node to judiciously chosen SIMD nodes; e.g., converge from 12 × 12 to 4 × 4, and link all 16 of these nodes to the MIMD node.

Link all five nodes in the 2 × 2 and the 1 × 1 layers to the MIMD node.

Ignore the remaining seven links to the MIMD node, or link them judiciously to lower layers of the pyramid.

(B′) Vary B above by linking 36 nodes from the 24 × 24 layer to the MIMD node. Now the additional 28 lines can be scattered at the higher layers of the pyramid.

(C) Combine nine MIMD nodes with one pyramid node:
Link 144 of the 218 lines from the nine 64-bit MIMD nodes to 144 of the 576 nodes of the 24 × 24 layer of the pyramid.

Link 64 lines to the 64 nodes in the 8 × 8 layer, 16 to the 4 × 4 layer, 4 to the 2 × 2 layer, and 1 to the 1 × 1 layer of the pyramid.

(D) Combine two MIMD nodes with eight pyramid nodes:
Link 8 of the 128 lines from the two 64-bit MIMD nodes to the apex of a different pyramid.

Put four of these pyramids so that their apexes are scattered through the 16 × 16 layer of a larger pyramid.

Put the other four pyramids so that their apexes form the 2 × 2 layer of the larger pyramid.

Thus this construction links eight subpyramids in a larger pyramid (which has all the usual pyramid links) to the two serial computers in the larger MIMD–SIMD network.

(E) An array or a set of arrays can be used to replace the pyramids in the above constructions.

(F) Instead of linking all MIMD nodes to the same pyramid, we can link several of these nodes to an array or to several different pyramids.

Many other variants are possible, with, for example:

(1) the nodes linked to nodes that are more widely scattered in the pyramid or array (because they come from a lower layer of a larger pyramid);

(2) the SIMD nodes linked to each MIMD node scattered and interspersed with the SIMD nodes linked to the other MIMD nodes;

(3) the MIMD nodes linked in ways designed to handle the shuffling of data effected by the buffer memories in the MPP (Batcher, 1980), or by banks of switches such as the flip network in the STARAN computer [24], the shuffle [25], and other networks [26], [27].

7. COMBINING SIMD PYRAMIDS AND ARRAYS WITH MIMD CONVENTIONAL COMPUTERS INTO A 50-NODE OPTIMAL (50, 7, 2) SINGLETON GRAPH

A substantially larger structure can be achieved by using the 50-node (7, 2) Singleton graph, in much the same ways as already illustrated for the 10-node (3, 2) Petersen graph. The 50 nodes can be partitioned into those allocated to pyramids, subpyramids, or arrays, and those allocated to MIMD network processors. The nodes allocated to pyramids can be linked to one or to several layers of the pyramid, scattered or packed as desired.

To give one simple example, a 7×7 array of 49 nodes can be linked directly to a scattered 7×7 array in a suitably larger array or layer of a pyramid, with one large serial computer as the 50th node.

A large variety of other systems can easily be designed to meet the mixture of requirements anticipated, along the lines suggested in the previous section.

8. POSSIBLE EXTENSIONS TO LARGER NETWORKS, BY COMPOUNDING GRAPHS

These still relatively small graphs can be compounded or combined into larger structures, using a variety of recently discovered combining operations that give relatively dense (but by no means optimal) graphs [28]. These compounded graphs have a two-level hierarchy:

(1) the basic graph (which forms a local cluster) and

(2) the larger graph that compounds a number of these basic graphs together. This gives the architect designing a multicomputer the opportunity to choose a basic cluster graph for whatever set of local characteristics are deemed important and also to link these basic graphs together in a relatively dense way.

Indeed, multicomputers like Cm*, which, have been designed with local clusters (usually some small number such as 5, 8, or 16 computers on a single bus or ring) linked to other local clusters (e.g., with a star, N-cube, or reconfiguring network), might be improved substantially by using a good compounding operation to tie together good local clusters. But this issue of compounding and constructing still larger graphs that specify the topology of a multicomputer network lies outside the scope of the present paper.

SUMMARY

The following comments summarize and expand upon the constructions given above.

1. Each node in the graph can be replaced by either an MIMD network, an entire multicomputer pyramid or array, a subpyramid or subarray, an individual MIMD computer, or a component of an individual computer.
2. Each bit-line to each MIMD computer can be linked to a different 1-bit computer in the SIMD pyramid, subpyramid, or array.
3. Not all of the lines need be used to link computers. Some can link a computer to an input, output, or mass memory device, when appropriate. Some need not be used at all.
4. The SIMD links can be concentrated in a single array or region, or they can be scattered about through the structure to which they are linked. Then information can be passed to them by moving it laterally from node to node in the array, or/and (if the larger structure is a pyramid) up through the log N interconnection links of the pyramid.

A few specific examples of such structures have been specified in this paper. The total set of possibilities is enormously larger, because each node in the original graph can be replaced by any type of node, and different types of nodes can be linked together in any desired way. Thus it becomes possible to design systems that combine different types of synchronous SIMD and asynchronous MIMD computers as most appropriate for particular kinds of programs.

REFERENCES

[1] Uhr, L. (1984). *Algorithm-Structured Computer Arrays and Networks: Parallel Architectures for Perception and Modelling*. Academic Press, New York.
[2] Flynn, M. (1972). Some computer organizations and their effectiveness, *IEEE Trans. Computers*, **21**, 948–960.
[3] Uhr, L. (1983). Pyramid multi-computer structures, and augmented pyramids. In *Computing Structures for Image Processing* (M. J. B. Duff, ed.). Academic Press, London, pp. 95–112.
[4] Uhr, L. (in press). Pyramid multi-computers, and extensions and augmentations. In *Algorithmically Specialized Computer Organizations* (D. Gannon, H. J. Siegel, L. Siegel, and L. Snyder, eds.). Academic Press, New York.
[5] Reddaway, S. F. (1978). DAP—a flexible number cruncher, *Proc. 1978 LASL Workshop on Vector and Parallel Processors*. Los Alamos, 233–234.
[6] Duff, M. J. B. (1978). Review of the CLIP image processing system, *Proc. AFIPS NCC*, 1055–1060.
[7] Batcher, K. E. (1980). Design of a massively parallel processor. *IEEE Trans. Computers*, **29**, 836–840.
[8] Dyer, C. R. (1982) Pyramid algorithms and machines. In *Multi-Computers and Image Processing* (K. Preston, Jr. and L. Uhr eds.). Academic Press, New York, pp. 409–420.

[9] Tanimoto, S. L. (1981). Towards hierarchical cellular logic: design considerations for pyramid machines. Computer Sci. Dept. Tech. Rept. 81-02-01, University of Washington, Seattle.

[10] Uhr, L. (1981). Converging pyramids of arrays. In *Proc. Workshop on Computer Architecture for Pattern Analysis and Image Data Base Management*, IEEE Computer Society Press, pp. 31–34.

[11] Siegel, H. J. PASM: A reconfigurable multimicrocomputer system for image processing. In *Languages and Architectures for Image processing* (M. J. B. Duff and S. Levialdi eds.). Academic Press, London.

[12] Briggs, F., Fu, S., Hwang, K., and Patel, J. (1979). PM4—a reconfigurable multimicroprocessor system for pattern recognition and image processing. *Proc. AFIPS NCC*, 255–265.

[13] Mago, G. A. (1980). A cellular computer architecture for functional programming, *Proc. COMPCON Spring 1980*, 179–187.

[14] Despain, A. M. and Patterson, D. A. (1978). X-tree: A tree structured multi-processor computer architecture. *Proc. Fifth Annual Symp. on Computer Arch.*, April, 144–151.

[15] Swan, R. J., Fuller, S. H. and Siewiorek, D. P. Cm*—A modular, multimicroprocessor. *Proc. AFIPS NCC*, 637–663.

[16] Wittie, L. D. (1978). MICRONET: A reconfigurable microcomputer network for distributed systems research. *Simulation*, **31** 145–153.

[17] Manara, R. and Stringa, L. (1981). The EMMA system: An industrial experience on a multi-processor. In *Languages and Architectures for Image Processing* (M. J. B. Duff and S. Levialdi eds.). Academic Press, London.

[18] Rieger, C., Bane, J., and Trigg, R. (1980). A highly parallel multiprocessor. *Proc. IEEE Workshop on Picture Data Description and Management*, 298–304.

[19] Cook, R., Finkel, R., Gerber, B., DeWitt, D., and Landweber, L. (1983). The crystal nuggetmaster, *Computer Sci. Dept. Tech. Rept.* 500, Univ. of Wisconsin, Madison.

[20] Lipovski, G. J., and Tripathi, A. (1977). A reconfigurable varistructure array processor. *Proc. 1977 Int. Conf. on Parallel Processing*, 165–174.

[21] Uhr. L. (1982). Comparing serial computers, arrays and networks using measures of "active resources." *IEEE Trans. Computers*, **30**, 1022–1025.

[22] Hoffman, A. J. and Singleton, R. R. (1960). On Moore graphs with diameter 2 and 3. *IBM J. Res. Devel.*, **4**, 497–504.

[23] Bondy, J. A. and Murty, U. S. R. (1976). *Graph Theory with Applications*, Elsevier, New York.

[24] Batcher, K. E. (1976). The flip network in STARAN. *Proc. 1976 Int. Conf. on Parallel Processing*, 65–71.

[25] Stone, H. S. (1971). Parallel processing with the perfect shuffle. *IEEE Trans. Computers*, **20**, 153–161.

[26] Siegel, H. J. (1979). A model of SIMD machines and a comparison of various interconnection networks. *IEEE Trans. Computers*, **28**, 907–917.

[27] Benes, V. E. (1965). *Mathematical Theory of Connecting Networks and Telephone Traffic*, Academic Press, New York.

[28] Bermond, J. C., Delorme, C., and Quisquater, J. J. (1982). Tables of large graphs with given degree and diameter. *Info. Processing Letters*, **15**, 10–13.

Chapter Three

An Approach to the Iconic/Symbolic Interface

Steven L. Tanimoto

1. INTRODUCTION

Image processing techniques are often classified into either low-level or high-level types. The low-level types consist of transformations whose domains and ranges are all two-dimensional arrays. High-level techniques produce outputs in scalar, list, tree, graph, or tabular form, rather than image-array form [1], [2]. Parallel processing can usually be applied to operations at either level. Low-level transforms can usually be computed quickly by cellular arrays of processors. High-level operations can be computed by networks of general-purpose (Von Neumann-style) computers.

In the human visual system, neurophysiologists sometimes make a distinction between the retinotopic levels and the nonretinotopic levels of information processing. A retinotopic level is a layer of neurons in the visual pathway between the retina and area 17 of the striate cortex, in which the neighboring areas of the visual field are represented by neighboring neural material. That is, the general spatial structure of the visual field is preserved in the spatial structure of the neural layer. A nonretinotopic level is a brain area in which the neural signals are so semantically coded that the spatial correspondence with the visual field is almost entirely lost.

Our problem is to provide guidance as to how the low-level–high-level (corresponding to the retinotopic–nonretinotopic) gap should be bridged in machine vision systems. Designing an interface between a cellular array processor and a team of general-purpose computers is challenging if one wants to make the data transfer speed compatible with the speeds of the array and the collection of general-purpose computers.

INTEGRATED TECHNOLOGY FOR
PARALLEL IMAGE PROCESSING
ISBN 0-12-444820-8

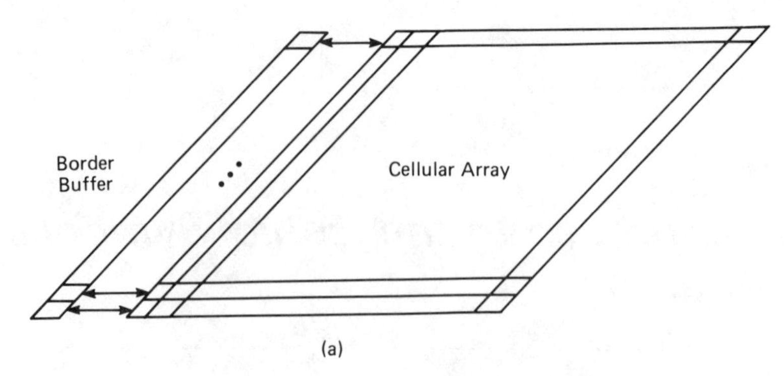

Fig. 1(a) Cellular array with border buffer

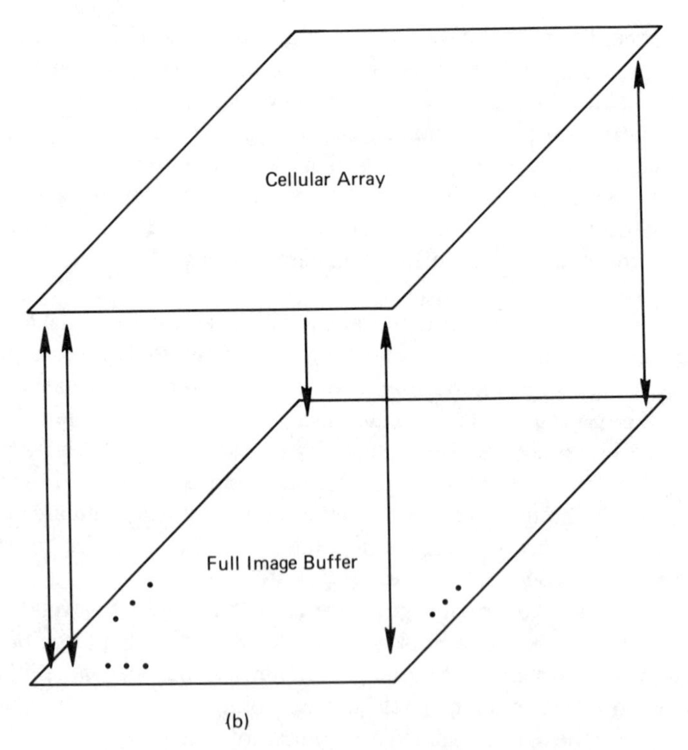

Fig. 1(b) Cellular array with full image buffer

2. IMAGE BUFFERS

Generally, a cellular array is loaded and unloaded through buffers that hold one or more rows or columns of image data. A border buffer is connected to all the cells along one side (or more) of the array, and all the cells on that side can load or unload values at once [see Fig. 1(a)].

A parallel-loading full-image buffer consists of an array of storage cells equal in number to that of the cellular processor. Once the buffer contains all the data to be input to the processor, the cells of the processor can all be loaded in one parallel operation [see Fig. 1(b)].

Both the CLIP4 [3] and the MPP [4] have parallel-loading capabilities, although the full-image buffers themselves are generally loaded serially. Thus, if a host prepares data (to be input to the cellular array) well in advance of the time the data are needed by the array, there need be no waiting by the array. However, if there are frequent I/O operations to and from the array, the buffer access by the host becomes a bottleneck, and the cellular array processing elements may be idle.

To provide a high bandwidth channel between a cellular array and several general-purpose computers, a more powerful interface is needed. Our answer is to propose a multiple-buffer method.

3. ICONIC/SYMBOLIC INTERFACE ARCHITECTURE

A straightforward organization of multiple I/O buffers to a cellular array is now described. A set of k full-image I/O buffers is piled into a stack, k high. The cellular processor sits on top of this pile and is connected through many parallel vertical lines to the pile of buffers. Each buffer is composed of special memory modules, called *hetero-ported memory modules* (see Fig. 2). With these modules, pixel data can move vertically between the cellular array and the storage area or horizontally between a general-purpose processor and the storage area.

When data move vertically, they move between the cellular array and one of the k image buffers in the pile. An entire image is transferred in parallel at once. When data move horizontally between a von Neumann processor and the storage area, they appear to be moving between an addressed memory location and the processor. Data move horizontally only a word at a time in any given level. However, independent transfers can take place simultaneously in different levels.

When data are to be transferred vertically, a selection is made that connects one of the levels to the cellular array. When data are to be transferred horizontally, the processor connected to the level of interest issues addresses (the image buffer forms part of the processor's address space).

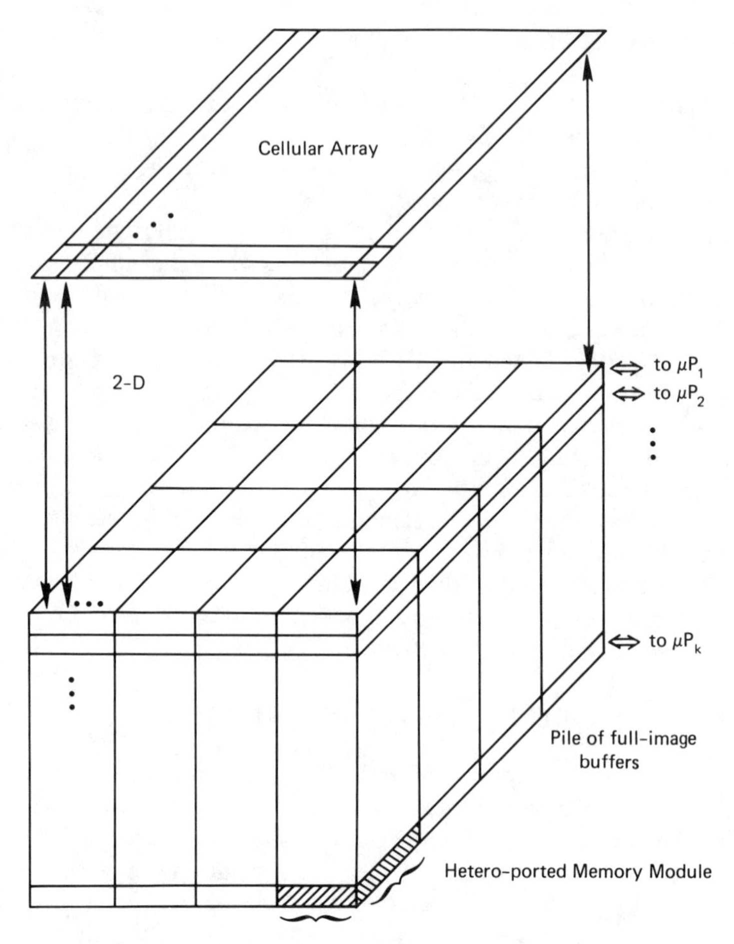

Fig. 2 Proposed architecture for the iconic/symbolic interface

4. HETERO-PORTED MEMORY MODULES

The memory modules described above are not commonly available. Here we describe the requirements for such modules.

A hetero-ported memory module (HPMM) is an array of storage elements grouped together for physical convenience and used as a component in building a pile of image buffers. It is characterized by the following parameters:

1. *Subimage width* (the number of columns in the subimages handled by the module)

2. *Subimage height* (the number of rows in the subimages handled by the module)
3. *Word size* (the number of bits that are simultaneously accessed in horizontal mode)
4. *Number of levels* (the number of levels of the pile of buffers to which the module contributes)
5. *Vertical multiplexing factor* (fully parallel vertical transfers are too costly in many applications, and some degree of multiplexing is desired)

There are potentially a large number of signal wires emanating from each HPMM. These are described as follows:

1. *Vertical I/O connections.* The number of these is equal to the product of the subimage width times the subimage height, divided by the vertical multiplexing factor.
2. *Horizontal I/O data lines.* The number of lines is equal to the number of levels times the word size.
3. *Horizontal address lines.* Here the log (base two) of the subimage size is the number of lines required. The subimage size is the product of the subimage width and the subimage height.
4. *Vertical level and word-bit selection lines.* During a vertical transfer, only one bit plane of a HPMM is accessed at a time. Log (base two) of the number of levels is the number of lines needed to select the level. In addition log (base two) lines are needed to select the bit within each word.
5. *Phase.* With multiplexed vertical transfers, some scheme for synchronizing each HPMM with the cellular array is necessary. One way to do this is to use a number of lines equal to log (base two) of the vertical multiplexing factor, to communicate the index of the current phase at all times.

5. THE SCHIZO CHIP

The HPMM could be realized as a chip. Here, specific parameters are proposed, and the name given to the proposed design is *schizo chip*, in recognition of the split personality of the HPMM.

Figure 3 shows the inputs and outputs to the schizo chip. To make best use of the chip area, parts of two separate image buffers, A and B, are stored on each chip. Each partial image buffer is 8 bits deep and represents a 16 × 16 subimage of the corresponding whole. A and B each have 8 horizontal address lines and 8 horizontal bidirectional data lines. Thus, there are 32 lines for horizontal I/O. For vertical I/O there are 32 bidirectional data lines, each multiplexed 8 ways, so that 256 bits can be transferred in 8 array cycles.

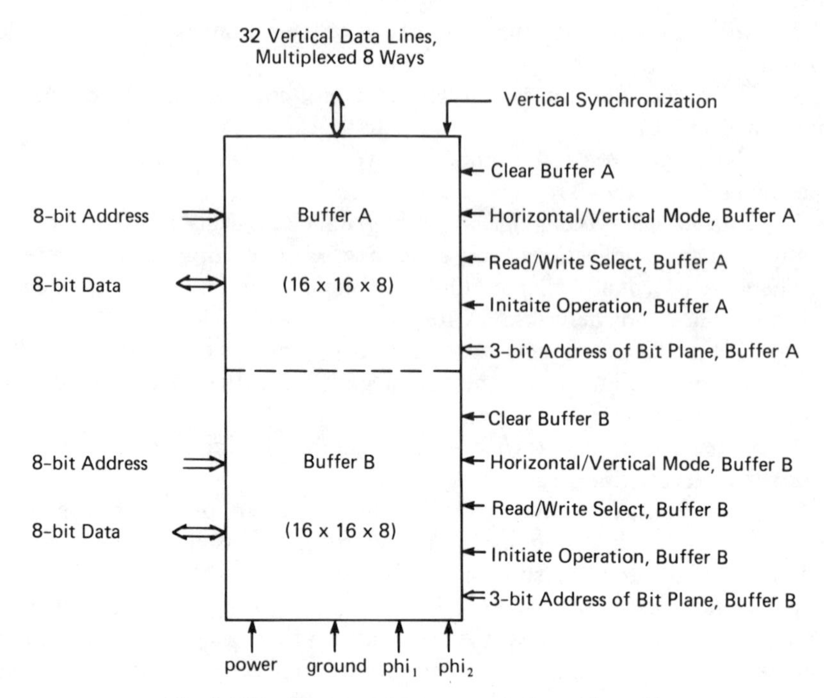

Fig. 3 The inputs and outputs of the schizo chip

A and B each have 3 bit-plane-address lines for selecting which bit plane is accessed in vertical mode.

A and B each also have the following lines: Read/write select, initiate I/O, clear entire buffer, and horizontal/vertical mode select. Additional pins on the chip are required for: vertical mode synchronization, power, ground, phi_1, and phi_2 (nonoverlapping two-phase clock). The schizo chip requires a total of 83 pins.

To load a binary image (vertically) into a bit-plane of buffer A in the schizo chip, the following must be done:

1. Select vertical mode on buffer A.
2. Select the *write* operation.
3. Present the 3-bit vertical address of the bit-plane desired.
4. Initiate the operation.

The chip then demultiplexes the signals on each of the 32 vertical data lines, using the vertical synchronization input as a clock. After 8 cycles, all 256 bits for the 16 × 16 subimage are stored.

For the microprocessor attached to buffer A to read a pixel value from buffer A (horizontally), the following are done:

1. Select horizontal mode on buffer A.
2. Select the *read* operation.
3. Present the 8-bit address of the pixel desired.
4. Initiate the operation

The complete image buffer A is made up of many schizo chips. Additional addressing logic (to that on the chips) allows only one of the chips to initiate a read or write operation in buffer A. The 8 bits of data are then placed on the horizontal data bus for buffer A.

The actual number of bits stored on one schizo chip is not large: 256 bits/plane × 16 planes = 4096 bits. However, there is more logic on a schizo chip than in a typical 4K memory chip. With 83 I/O lines, a lot of chip area is taken up by pads.

To build an interface between one cellular array of size 128 × 128 and two von Neumann processors, an 8 × 8 array of schizo chips would be necessary, for a total of 64 schizo chips. If 16 rather than two von Neumann processors are to be used, then 512 schizo chips would be required.

6. DISCUSSION

The interface architecture proposed here permits one parallel-cellular array processor to cooperate with a collection of general-purpose processors in an intimate fashion without any bottlenecking at the interface. It is foreseen that such an arrangement would allow such functions as image acquisition and display to be carried out by independent processes running at their normal rates.

There is no theoretical limit to the number of buffers that could be put in the pile. The physical problem of providing all the vertical interconnections is somewhat reduced by multiplexing.

We have not addressed the problem of how two or more parallel cellular arrays could be interfaced for high-bandwidth, asynchronous communication.

REFERENCES

[1] S. L. (1980). Image data structures. In *Structured Computer Vision: Machine Perception Through Hierarchical Computation Structures* (S. L. Tanimoto and A. Klinger eds.). Academic Press, New York, pp. 31–55.

[2] Tanimoto, S. L. (1976). An iconic/symbolic data structuring technique. In *Pattern Recognition and Artificial Intelligence* (Chen, ed.). Academic Press, New York, pp. 452–471.

[3] Duff, M. J. B. (1976). CLIP 4: A large scale integrated circuit array parallel processor. *Proc. Third International Joint Conference on Pattern Recognition*. Coronado, California, pp. 728–733.

[4] Batcher, K. E. (1983). Architecture of the MPP. *Proc. 1983 IEEE Computer Soc. Workshop on Computer Architecture for Pattern Analysis and Image Database Management*, Pasadena CA, Oct. 12-14, pp. 170–174.

Chapter Four

Multicluster: An MIMD System For Computer Vision

Anthony P. Reeves

1. INTRODUCTION

Computer vision applications frequently require very high computation rates for both low-level image processing and object identification stages, which can only be achieved with a parallel computer organization. Low-level image processing schemes are frequently characterized by simple deterministic algorithms applied to all elements of the input image. Many novel highly parallel SIMD architectures have been designed for this task which take advantage of the high functional complexity made available with VLSI technology. Algorithms for object identification are much less developed or standardized. Many of these algorithms are nondeterministic and may deal only with small regions of an image. Frequently an MIMD approach can be used; for example, to identify a set of segmented objects a separate concurrent task could be used for each object or even each object hypothesis.

Multicluster is a general MIMD framework for high-speed parallel computation. A system in this framework consists of a set of computational modules called *clusters*; each cluster contains one or more 32-bit microprocessors, high-speed arithmetic support, local memory, and an intercluster interface. The processor interconnection scheme will involve a hierarchy of at least two levels: the intracluster network, which will be determined by technological considerations and the intercluster network, which will be based more on algorithmic and general architectural considerations. The system should be fault tolerant to the failure of one or more clusters (or intercluster connections) and should suffer a graceful degradation of performance as clusters or parts of clusters fail. Futhermore, it should be possible to add new processors to a system as the demands on the system increase.

INTEGRATED TECHNOLOGY FOR
PARALLEL IMAGE PROCESSING
ISBN 0-12-444820-8

A Multicluster system is suitable for the parallel implementation of all computer vision algorithms. Special purpose SIMD processors can be attached to clusters for more cost-effective, low-level image processing and the implementation of other important well-structured algorithms. In a general sense, Multicluster embodies a hierarchical organization of three fundamental parallel processing architectural types: loosely coupled MIMD, tightly coupled MIMD, and SIMD.

The design of such a system involves open research questions concerning hardware organization, task allocation, and programming environment. Our approach is first to consider a hardware foundation that is capable of efficiently implementing a set of representative algorithms. Initial research for the Multicluster project is centered on a hardware multiprocessor testbed system based on Intel 432 32-bit microprocessor components. Work is in progress to determine the intercluster network that is needed to support various algorithms.

The remainder of this chapter is divided into three main sections. First, the Multicluster hardware framework is described. A major hardware design problem, the organization of the global interconnection network, is outlined. Second, software techniques for implementing algorithms on the Multicluster framework are considered. Finally, a simple near-neighbor primitive algorithm is analyzed to determine global network design parameters.

2. THE MULTICLUSTER FRAMEWORK

2.1 Multi-microprocessor organization

A convenient way to organize a multi-microprocessor system is so that all processors share, and have equal access to, the same memory system. The address space is identical for all processors; this simplifies the organization of the programs in general. However, conflicts that can cause a significant decrease in overall performance can arise when two or more processors attempt to modify the same memory location simultaneously.

One way to implement the above scheme is to organize the memory in modules and to interconnect processors and memory modules through a crossbar interconnection network that can independently connect any processor to any memory module. The multiprocessor C.mmp [1] is organized in this way. There are three basic drawbacks to this scheme. First, the crossbar switch is costly, especially when the number of processors is large. Second, an arbitration mechanism is needed for processors attempting to access the same module simultaneously. Finally, all memory accesses are made through the interconnection network, which slows down all processor operations. However, cache memories can be used to alleviate this problem.

In a recent design for a large scale system of this type, called the NYU Ultracomputer [2], a simpler but more restricted omega network is proposed. This network is pipelined, involves message switching, and has queues in each node to maintain a high bandwidth. It also involves an interesting hardware mechanism for dealing with simultaneous accesses to the same memory location. The potential problems with this network are a high memory latency which is statistical in nature, and fault tolerance.

A second approach, embodied in the Cm* design [3], is to associate local memories with each processor and use a hierarchical multiple-bus system to interconnect processors. In Cm*, processors are connected in clusters; each cluster contains a maximum of 14 processors with local memories. Clusters are interconnected with a second bus system. The address space appears to be continuous to each processor; however, local accesses are achieved in 3.5 μs, intracluster accesses require 9.3 μs, and intercluster accesses require 26 μs. The advantage of this organization is that many more processors can be used with a reasonable interconnection cost; and fast, full-speed accesses can be made to local memories. The disadvantage is that remote data accesses are slow, and the processor is idle while these accesses are made. The key to effective use of this system is to minimize the number of remote accesses; fortunately, due to the local properties of many algorithms, it is frequently possible to ensure that most memory accesses (over 90%) are to local memories.

The idle time that occurs when a processor has to make a remote access can be dealt with in two ways. First, one can use a multitasking mechanism, which performs an automatic context switch when a remote access is made. Such rapid context switching is a feature of the HEP processor [4] which interleaves the execution of several tasks at the instruction level to achieve high throughput. This scheme has two disadvantages in the present context; first, a higher degree of problem fragmentation is necessary to achieve high performance, and second, such task switching would require larger cache memories which can be a substantial part of the processor cost.

The second method is to separate the operations of local and remote access. The processor is constrained to access only local memory whereas an interconnection hardware device is responsible for interprocessor data transfers. The disadvantage of this method is that the memory no longer has the appearance of a single shared unit because remote date accesses must be treated like I/O transfers rather than memory accesses. Remote accesses usually require several instructions to set up; however, a block of words can be transferred to local memory by DMA while the processor executes a different part of the task. The ZMOB processor [5], which consists of 256

Z80 microprocessors, is organized in this way with a high-speed ring interconnection network. A data transfer between any two processors is guaranteed within 25.7 μs (if no conflicts occur) whereas a local memory access requires only 350 ns.

Systems of this type are classified as *loosely coupled* and are similar to microprocessor local networks. However, there is a major difference in the design goals of the two types of system. In a computer network the usual goal is to share a number of resources amongst a large number of users. The key here is that the work load is typically a large number of relatively independent tasks. Frequently non-numeric applications such as word processing and data base management are important, whereas the objective of the systems considered here is to improve the throughput performance for a single scientific task. Therefore, though similarities occur in the appearance of the hardware organization, significant differences are likely in the details of the implementation of the two system types.

2.2 Design goals

The general design goal of Multicluster is to achieve a cost-effective, high-speed MIMD system for scientific applications. There are many formidable problems in designing an MIMD system, including subtask allocation strategy and interconnection network design. However, a set of ideal characteristics for a general purpose MIMD system can be defined:

(1) *Extensibility*. It should be possible to add processors as demands on the system increase, just as one now adds additional memory to processors.

(2) *Fault tolerance*. It should be possible to reconfigure the system if any component becomes faulty so that computation can continue with a minimum of interruption. Furthermore, the reduction in performance should be proportional to the contribution made by the failed processor, and no recompilation of currently executing tasks should be necessary.

(3) *Programmability*. The user of the system should be able to write programs in a high-level language without having a detailed knowledge of the architecture of the system and without knowing how many processors the system has. Programs developed in this language should be both dynamically extensible and fault tolerant to take advantage of the features mentioned above.

2.3 Multicluster architecture

The general framework of the Multicluster system is outlined in this section. The system may be considered as a set of computing modules or processor cluster (PCs) interconnected by a global interconnection network

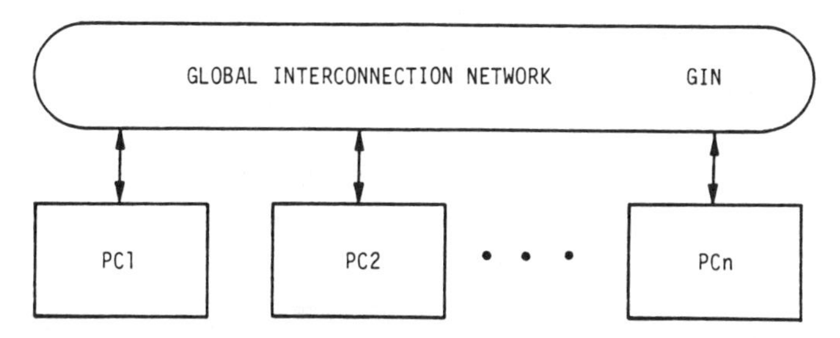

Fig. 1 Multicluster general organization

(GIN), as illustrated in Fig. 1. The detailed design of the GIN depends on many factors, some of which are listed below:

(1) Number of PCs
(2) Algorithms to be implemented and their distribution among the PCs
(3) Processing speed of the PCs
(4) Fault tolerance requirements.

Two main design parameters are topology and speed. Although these are interrelated, the topolgy must match the data flow of the desired algorithms; and the speed, determined by the technology used and the degree of parallelism in the network, must match the processing speed of the PCs. Many different topologies are potential candidates for the GIN. At one extreme, the full crossbar organization is likely to be too expensive whereas at the other extreme the single ring scheme is probably too slow and is not very fault tolerant. Between these extremes are several alternatives such as multibus networks, permutation (n log n) networks, hierarchical (nonhomogeneous) networks, and hybirds formed from more than one network type. Factors that affect the selection of a network include the available technology, the speed of the PCs and the performance for benchmark algorithms.

The organization of a cluster of processors for Multicluster is shown in Fig. 2. The cluster consists of four classes of devices connected to a local interconnection network (LIN). A serial processor (SP) and memory module (MM) together with the LIN form the heart of the PC and can be constructed with conventional microprocessor components. The LIN will probably be a bus system, depending on the exact microprocessor that is selected for the SP and the least expensive organization available. The MM is to be constructed with conventional microprocessor memory technology. The microprocessor will have a local cache memory. There may be more than one SP in a PC; for example, the Intel system 432/600 [6] permits up to five SPs to be connected to a single LIN.

Other devices to be connected to the LIN include I/O processors (IOP) and attached processors (AP). More than one IOP or AP may be connected to

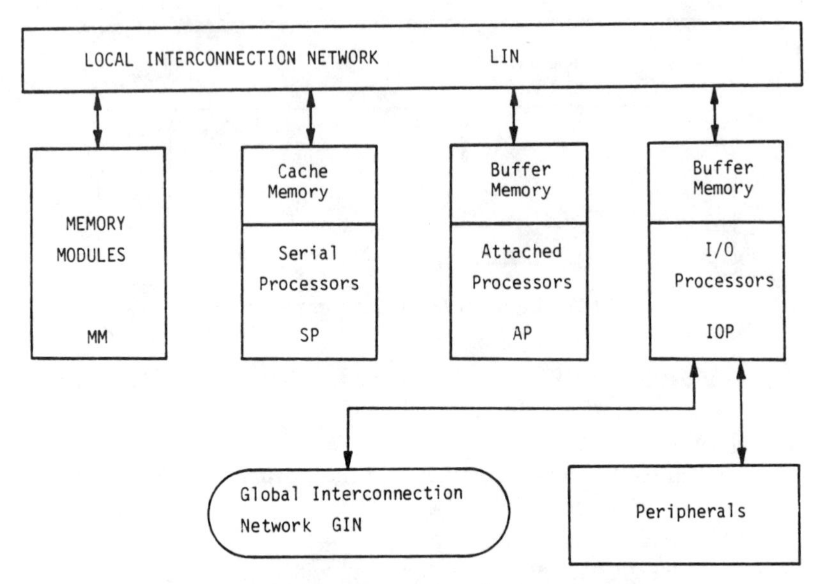

Fig. 2 Multicluster processor cluster organization

the LIN to meet the performance requirements of the PC. I/O processors perform two functions: communications with peripheral devices and communications with the GIN. They can contain buffer memories to deal with high speed or congested devices on interconnection networks.

The APs are special-purpose processors that can be used to enhance performance for specific applications. For example, as the floating-point performance of current VLSI microprocessors is not yet very good, a fast floating-point processor may be a very useful additon to the PC. Other possibilities for APs are vector pipelines, systolic arrays, and small processor arrays. There is no need for all PCs to have the same configuration; some PCs can be considered to be "experts" for certain applications by virtue of their APs.

2.4 Processor cluster design

The number of serial processors in a cluster will, in general, be a function of the available components rather than a specific design parameter. For conventional processor and bus systems this number can be one or two. However, a greater number of processors in a cluster is desirable for two main reasons. First, for fault tolerance it is useful to have several processors, so that if one fails, the other resources in the cluster, including memory content, will not be lost. Second, it is simpler to organize an efficient program in the true shared-memory environment that exists in a cluster.

An important feature of the PC design (and of all shared-memory systems) is to have a good cache memory associated with each procesor. In additon to providing fast local memory access for the processors, the cache memories also reduce the load on the LIN, which allows more processors to be connected to a cluster.

The Intel 432 design group has developed an architecture that is most appealing for this application. The 432/600 development system permits several processors to operate on a single bus system concurrently; the number of processors can be made transparent to the programmer. With the more recent 432 chips [7], a set of processors can be connected to a set of memory modules via a crossbar-like interconnection network. The nodes that interconnect the data buses are functionally complex chips that can implement a variety of different memory mapping schemes and provide good fault tolerance.

This design achieves the goal of graceful degradation. A cluster, in this context, consists of three fundamental components—processors, memory modules, and data buses. If any single component fails in a cluster containing a multiplicity of all three component types, then the cluster can be reconfigured so that there is only a relatively small decrease in performance or capabilities.

2.5 Interconnection network design

The design of the GIN is the main challenge to the architect of a Multicluster system; the design of each cluster for a given microprocessor type is more straightforward. The GIN must be matched to the interprocessor data flow which is algorithm dependent. It should also be fault tolerant.

The speed of data transfers will be determined by the technology used and the width of the data paths. For fault tolerance, a network based on single wire links, such as Ethernet, may be considered. This type of network would have the most independent links between clusters and the most graceful fault tolerance. Unfortunately, this network would also have the worst data transfer speed because the data channel is only one-bit wide. If multiple wires are used then there should be eight or more to a channel because there must be enough so that some wires can be used for control and fault tolerance and still have multiple data wires for an adequate data bandwidth.

Since a full crossbar is too expensive to implement, the topology of the network must be matched to the data flow of the algorithms to be implemented. However, this data flow is very difficult to know in advance because many MIMD algorithms are nondeterministic. Furthermore, algorithms can frequently be changed to achieve the best results with a given network.

The global interconnection network design is probably the most difficult hardware design task in an MIMD system. The suitability of a particular network design, or the design of most other MIMD system components, needs to be tested by performance analysis. The following section outlines a hardware multiprocessor testbed system that will be used for testing designs through simulation. Performance analysis of a Multicluster system is considered in Section 5.

3. MULTICLUSTER TESTBED SYSTEM

A hardware multiprocessor testbed system is being developed at Cornell University, with the assistance of Intel, to conduct research with the Multicluster framework. This system is based on the Intel 432 32-bit microprocessor, which is programmed in Ada.

The organization of the multiprocessor testbed is shown in Fig. 3. The main processing is achieved with two 432/670 computer systems. Each 432/670 is a single-bus multiprocessor system that can contain two I/O processors and up to four data processors. These two systems are connected through the multibus of the link processor (LP). The link processor is an 8086-based system that will be used to control and monitor the link between

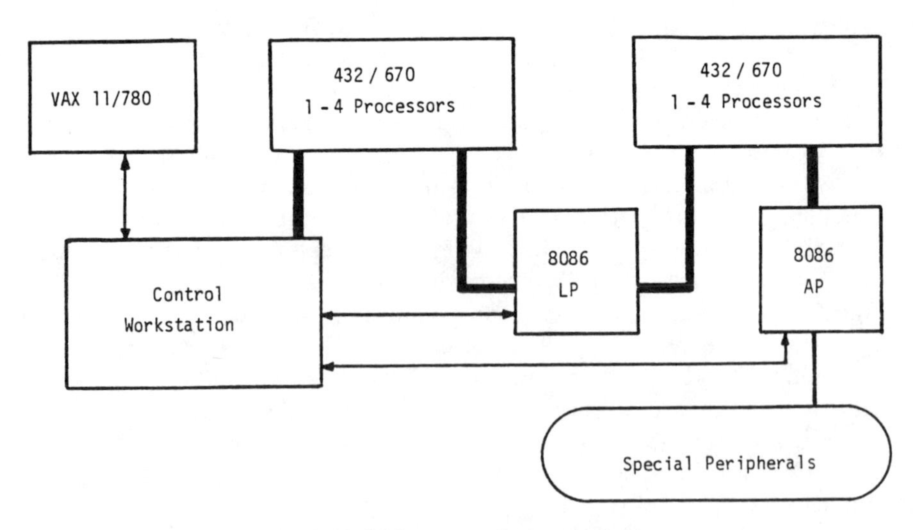

Fig. 3 Multicluster testbed organization

the two 432/670 systems. Programs for this testbed will be developed on a VAX 11/780 in Ada. The control work station will transfer compiled programs to the testbed processors and control the execution of programs on the testbed. The attached processor (AP) is a second 8086-based multibus system that will be used to communicate with special peripherals, such as an image display device, that need direct access to the testbed processors.

Each 432/670 will be used to simulate one or several clusters, and the LP will be used to simulate the global interconnection network. This testbed will offer two unique opportunities for research. First, the 432/670 is a true multiprocessor system and is commercially available. Software support makes the number of data processors in the system transparent to the user; programs automatically use all the available processors. Therefore, it is simple with this system to experiment with parallel algorithms and to measure the effect of different numbers of processors in a cluster. Second, Ada, the programming language of the system, is one of the few high-level programming languages that includes tasking capabilities in addition to a large range of features for scientific applications. This will provide an opportunity to investigate the suitability of Ada for the Multicluster framework. It will also allow software to be developed and tested for a full-sized multicluster system.

4. ALGORITHMS AND ARCHITECTURES

An early belief was that parallel computer architectures should be designed to implement efficiently the algorithm used in a given application. Since few parallel systems have been built, a more common approach has been to investigate how a particular application can be implemented on a given architecture. In many applications the algorithms are not fixed entities, and reformulation of serial algorithms can lead to a simpler, more effective implementation. A major problem that has been observed is that it is difficult to get the users to formulate a problem for a parallel architecture. But "thinking parallel" is essential, because it is impossible now (and probably in the foreseeable future) for vectorizing compilers to *change* the algorithm for a faster implementation, giving perhaps slightly different but equivalent results. An example of this type of algorithm development is the solution of PDEs [8]. The successive over-relaxation method has good convergence properties when compared with the Jacobi method but cannot be directly implemented in parallel. However, by using odd-even ordering, an effective parallel algorithm can be developed for most parallel processors. These algorithms perform different operations but converge to similar results. The algorithm changes require an in-depth understanding of the mathematical model being used, which is the domain of the application programmer.

The transition to "thinking parallel," though perceived as an obstacle to the use of current supercomputers and array processors, is only a small conceptual step. Most high-level languages for these systems are close to a matrix algebra, with which many users are already familiar for mathematical expressions. The parallel notation may frequently be much clearer than the equivalent sequential program: however, users' familiarity with serial processors and their languages makes the transition to a parallel notation a painful, though necessary, task.

This situation may soon get worse rather than better. The newer, MIMD, parallel architectures may have programming languages that have constructs dealing with asynchronous communication. Furthermore, the dichotomy between local and global memory access times may become an important algorithmic consideration. New high-level language constructs will be needed for such systems, but also new temporal concepts of algorithm design need to be understood by the applications programmer.

An example of this type is the implementation of a sorting algorithm on Cm* [9]. Cm* has a significant dichotomy between local and global memory access times, especially when many global accesses occur. The algorithm analyzed for Cm* is a modified version of quicksort (an algorithm that is considered one of the best for serial processors). This algorithm has a theoretical processor utilization on 20,480 data elements of about 50% for 10 processors, reducing to about 15% for 50 processors. However, in practice, much lower utilization resulted from global memory conflicts.

An alternative parallel algorithm is as follows:

1. Distribute equal numbers of data elements to each processor.
2. Compute coarse statistics on random samples of the data.
3. Split the expected range so that each processor is allocated a subrange with the same number of elements estimated from the statistics.
4. Each processor distributes the data, in packets, to respective subrange processors.
5. Each processor sorts its subrange with quicksort.

On a serial computer, steps 1–4 are a waste of time. In the parallel case, we reorganize the data so that the intensive sorting operations in step 5 are all local. It is possible, if the statistical estimates are good, and if steps 1–4 take much less time than step 5 (which should be true for large data sets) that a processor utilization greater than 50% may be expected.

Further refinements may be possible. For example, if some processors have disproportionally large numbers of data elements to sort in step 5, then a further partitioning of the data may be advantageous, sending half the data to the processors currently having the least to sort. Though such modifications are conceptually simple, their expression in current high-level languages for an asynchronous multiprocessor system are overly complex.

4.1 Locality transforms

We need to avoid two fundamental data-flow patterns to achieve efficient algorithms. The first is a large amount of interprocessor transfers that overload the interconnection network, e.g., the sorting experiment on Cm*. The second data flow to be avoided is having a single memory or processor that interchanges data with a large number of processors, because this may result in a local bottleneck even though there is plenty of interprocessor bandwidth.

The concept of the locality transform is introduced to develop algorithms without the above problems. A locality transform is an interprocessor data mapping that distributes data among the processors so that the interprocessor data transfers for subsequent processing are minimized.

Much of the art in programming on the MIMD system is in defining effective locality transforms. Examples of some locality transforms follow. The sorting algorithm mentioned previously uses data-dependent locality transforms to allocate data to the processors. A histogram algorithm has been described for ZMOB [5] which is in two stages: First, each processor calculates a local histogram, then the data counts are distributed to the processors so that each processor computes the total for one histogram bin. The very efficient ZMOB algorithm for distributing the subcounts is an excellent example of a locality transform. Finally, near-neighbor problems such as PDEs and low-level image processing are best stored so that each processor contains a consecutive subarray to minimize interprocessor communication.

In many areas we can define at least two phases of computation with different data structures, for which an intermediate locality transform could be developed. Consider generating a display in computer graphics, for example. Objects or surfaces could be distributed among the processors for perspective transformations. However, for a two-dimensional image generation a more convenient distribution might have subarrays of pixels associated with each processor. A locality transfer could be used to broadcast the object descriptions to the pertinent pixel processors.

4.2 Macro data flow

One way to reduce the effects of high interprocessor latency is to transfer data in complete blocks or structures rather than individual data elements. Furthermore, since programs are read only, they can be distributed to all clusters. That is, every cluster has a copy of all programs that may be run.

Therefore, there is never a need, during data processing, to transfer programs between clusters or processors.

In this context macro data-flow techniques may be considered. For this mode of operation, a data structure is assembled consisting of all the data for the execution of a subroutine, for example, plus a key that identifies the subroutine to be executed. In this way a busy cluster can assemble these data structures and distribute them to the currently least-busy clusters for processing. A result data structure (with a key indicating how the results should be processed) is returned to the initiating cluster. In a more complex scheme, data structures can be dispatched from different clusters to rendezvous at a third cluster for processing.

4.3 Macro pipelining

Macro pipelining has been proposed by several researches as a method of distributing large tasks on a distributed system. This technique can be used when the data are in the form of large arrays in which similar operations are to be performed on each element. A set of dedicated channels are set up between a set of processors, and the data flow through the processors in a pipeline manner. Unlike the conventional pipeline, the functions implemented by each processor stage can be arbitrarily complex.

The advantage of this scheme is that it is simple to set up, and it can be efficient in applications such as image processing in which the I/O bandwidth is very high [10] or when a large number of programs are running in each processor [11]. The problems with this scheme are that the system is constrained to the speed of the slowest processor function, links must be permanently established between processors, and the number of processors used must equal the number of pipeline stages, which means that any additional available processors will be idle. It seems that while macro pipelining may be useful in some special applications, it will not be very useful for more general problems.

5. PERFORMANCE ANALYSIS

A classical approach to the problem of performance analysis on an MIMD system is the use of a statistical model for an algorithm and the system. Obtaining reliable results for large systems is very difficult with this method. Frequently, computer vision algorithms possess a large degree of determinism; this may be difficult to model statistically but opens the possibility of more direct methods.

The following analysis is for a completely deterministic problem that results in bounds for the maximum latency between clusters to achieve a given processor utilization.

5.1 A simple Multicluster performance model

A simple Multicluster performance model has been developed to estimate system parameters for different tasks. A Multicluster system is represented by a 6-tuple,

$$MC = \{N, M; P, t_c, t_l, t_t\}$$

where

N is the number of data elements to be processed,

M is the number of clusters.

P is the number of processors in a cluster,

t_c is the time required for an elemental operation, i.e., the time for a single processor to do one computation (including local memory access where necessary),

t_l is the latency required to set up a data transfer between clusters,

t_t is the time to transfer an element between clusters.

A cluster is characterized by the parameters P and t_c. The simple model lumps together local memory access time and processing time into t_c. The global interconnection network is characterized by t_t and t_l. In general, data will be transferred in blocks; the time to transfer a block of k items is $t_l + kt_t$. The simple model assumes that the latency and transfer times between all clusters are identical and that there is no blocking. It is possible to compute a lower bound on the performance of a system by using the largest possible latency and transfer times for t_l and t_t. Significantly better results could obtained for some interconnection schemes by using a more complex model.

Our interest is to explore how the global network parameters t_l and t_t affect the performance of a Multicluster system implementing an algorithm. Two additional parameters are used to indicate critical latency values; these are defined as follows:

t_{lz}, the zero-equivalent latency, is the maximum latency that can be tolerated without any decrease in performance; i.e., the same performance as for $t_l = 0$ is achieved.

t_{lh}, the half-performance latency, is the maximum latency that can be tolerated without decreasing the performance of the system by more than 50%.

For algorithms in which data transfers cannot be overlapped with processing, t_{1z} will be zero. In these cases t_{1h} gives a measure of the impact of latency on performance. Other fractional latencies can be defined when appropriate.

5.2 A near-neighbor task

The task to be studied is the computation of a near-neighbor operation for a matrix in which each element is replaced by a function of itself and its eight near neighbors. This operation is fundamental to many low-level computer vision tasks such as filtering, edge detection, and relaxation.

The matrix is distributed among the clusters, each cluster containing a contiguous square region. For a cluster to compute the results for its submatrix it must access data from adjacent clusters. This is illustrated in Fig. 4. The submatrix in a cluster is considered as a center region A and eight 1-element-wide edge segments B–I. Eight surrounding 1-element-wide segments (a–h) are required from eight other clusters to compute results for the eight edge segments. The algorithm proceeds as follows: while A is being processed, a is being transferred from another cluster so that it is available for B to be processed. Data from other clusters are acquired as fast as possible in the order a–h while near-neighbor processing proceeds when data are available, in the order A–F. The processors will wait idle if the global network is not fast enough to keep at least one data element ahead of the processing.

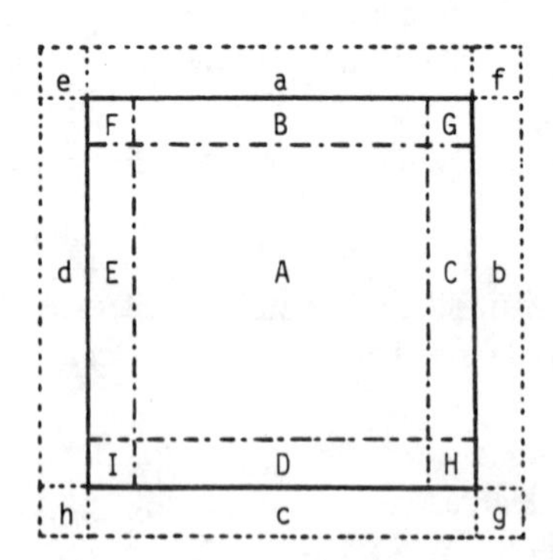

Fig. 4 Submatrix segments for near-neighbor task. A–I are data segments to be processed. Data segments a–h are to be obtained from other clusters.

The time T needed for a near-neighbor operation to compute is estimated from a function of the six model parameters

$$T = f(N,\ M,\ P,\ t_c,\ t_t,\ t_l).$$

Some example results of computing t_{lz} and t_{lh} for different model parameters are given in Figs. 5–7. In Fig. 5 the effect of the number of processors in a cluster on t_{lz} is shown, and t_c and t_t are both set to 10 μs. For this task t_t is the time to compute one complete result element when all the required data are locally available. Plots are shown for different n, which is the number of elements processed in each cluster, i.e.,

$$n = N/M$$

We can see from this graph that, for example, if $P = 8$ and $n = 1000$, then

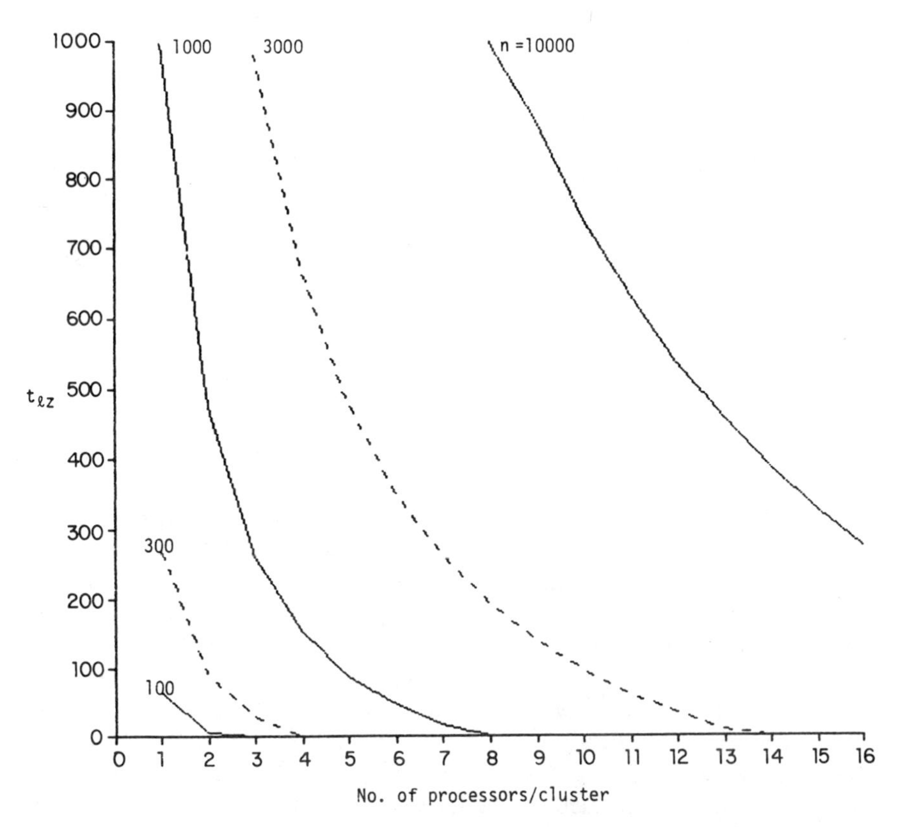

Fig. 5 Graph of t_{lz} in microseconds versus P for different n, with $t_t = 10\ \mu$s

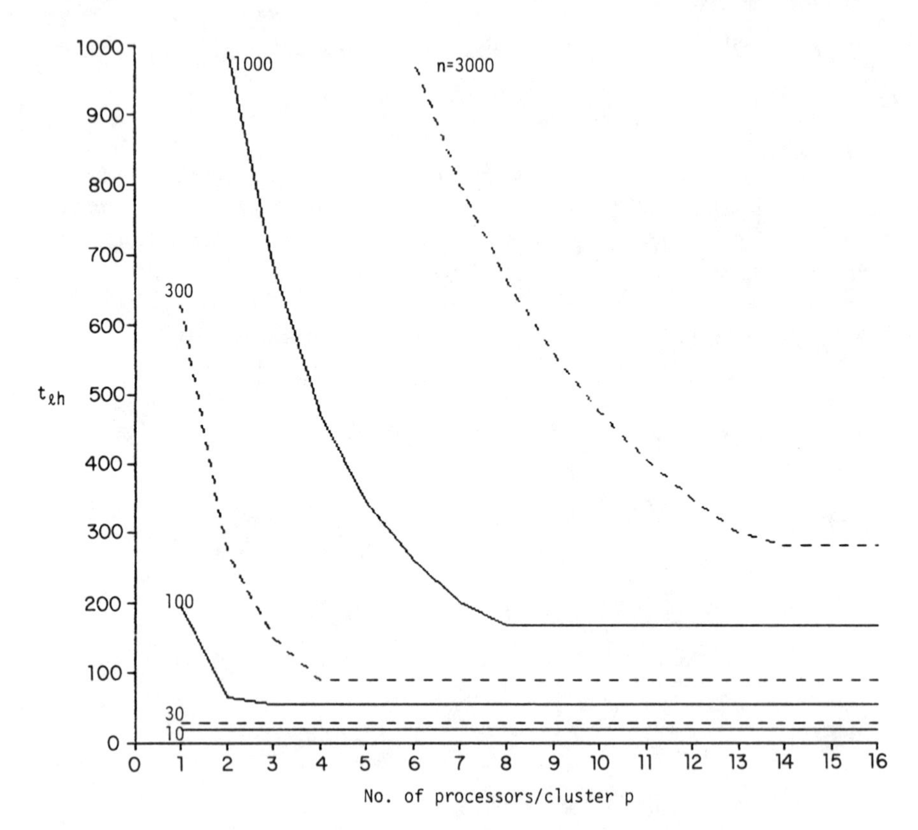

Fig. 6 Graph of t_{lh} in mucroseconds versus P for different values of n with $t_{\mathrm{t}} = 10\ \mu\mathrm{s}$

the performance will be degraded if t_{l} is greater than zero. However, with $P = 4$, a latency of 200 μs will cause no degradation in performance.

A similar graph but for t_{lh} is shown in Fig. 6. From this graph we can see that a latency of 200 μs with $P = 8$ and $n = 1000$ will cause a decrease in performance of 50%. A latency of over 400 μs causes a 50% performance decrease when $P = 4$ and $n = 1000$.

From Figs. 5 and 6 we can see that as n increases, t_{lz} also increases. This occurs because for the near-neighbor task the computation for each cluster is proportional to n^2, whereas the data to be transferred are proportional to n. In Fig. 7, t_{lz} is plotted versus n; plots for different values of t_{t} are given. P is set to 10, and t_{c} is 10 μs. From this graph we can see how small n must be before the latency degrades the system performance. That is, we mush have sufficient n to be to the right of the relevant plot line for the latency to have no effect. For example, with $t_{\mathrm{t}} = 1\ \mu$s and $t_{\mathrm{lz}} = 10\ \mu$s, n should be 150 or larger.

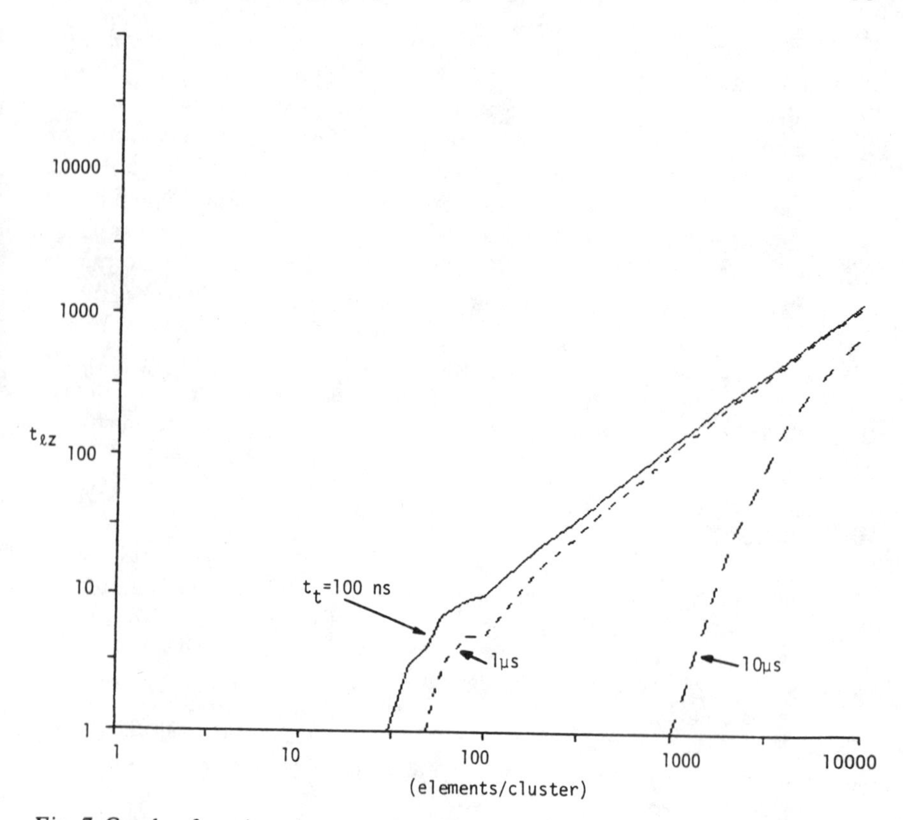

Fig. 7 Graph of t_{lz} in microseconds versus the number of elements/cluster for different t_t with the number of processors per cluster $p = 10$

6. CONCLUSIONS

The Multicluster MIMD organization for computer vision has been described. This framework embodies the concepts of a hierarchical memory system with fast local access and slower, usually block-oriented, access to more distant memory. Multicluster is currently oriented towards the 32-bit microprocessors that are available with todays technology. However, it is expected that these concepts will extend to future technologies that involve much higher levels of integration.

Many problems remain in designing an efficient system of the Multicluster type. A major problem at the hardware level is to design the global interconnection network. The throughput and data-flow requirements for this network need to be determined by the analysis of algorithms that are tc

run on the system. An example analysis for a simple important deterministic algorithm has been described. Two latency parameters t_{1z} and t_{1h} have been introduced, which are fundamental network requirements to achieve optimal performance and 50% of the optimal performance, respectively, for the given algorithm and system. These are also used to indicate how the global network requirements vary with other system parameters such as processing speed, number of processors per cluster, network data transfer time, and data size. Further work is needed in the areas of performance analysis of nondeterministic algorithms and the analysis of different network topologies.

A hardware multiprocessor testbed system has been described. This will be very important for developing software techniques and programs. It will also be useful for verifying, by simulation, the results of the performance analysis.

REFERENCES

[1] Wulf, W. A., and Bell C. G. (1972). C. mmp—A Multi-Mini-Processor, In *AFIPS Conference Prcceedings, FJCC* Vol. **14** Part II pp. 765–777.

[2] Gottlieb, A., Grishman, R., Kruskal, C. P., McAuliffe, K. P., Rudolph, L., and Snir, M. (1983). The NYU ultracomputer-designing an MIMD shared memory parallel computer. *IEEE Trans. on Computers* **C-32** (2) February.

[3] Fuller, S. H., Ousterhout, J. K., Raskin., L. Rubinfeld, P. L., Shindhu, P. J., and Swan, R. J. (1978). Multi-microprocessors: An overview and working example. *Proc. IEEE* **66**, 216–228.

[4] Smith, B. J. (1981). Architecture and application of the HEP multiprocessor computer system *SPIE Real-Time Signal Processing IV* **298**, 241–248.

[5] Rieger, C. (1981). ZMOB: Hardware from a user's viewpoint. *Proc. IEEE Conference on Pattern Recognition and Image Processing*, August.

[6] Intel Corporation (1981). *Intel System 432/600 System Reference Manual*, Order no. 172098-001, December.

[7] Intel Corporation (1983). *Electrical Specifications for iAPX 43204 Bus-Interface Unit (BIU) and iAPX 43205 Memory Control Unit (MCU)*. Order No. 172867-001, March.

[8] Hockney, R. W. and Jesshope C. R. (1981). *Parallel Computers*. Adam Hilger Ltd, Bristol.

[9] Deminet, J. (1982). Experience with multiprocessor algorithms, *IEEE Trans. on Computers* **C-31**(4), 278–288.

[10] Luetjen, K., Gemmar, P., and Ischen, H. (1980). FLIP: A flexible multi-processor system for image processing. *Proceedings of the Fifth International Conference on Pattern Recognition*. pp. 326–328.

[11] Bruner, J. D., and Reeves, A. P. (1982). An image processing system with computer network distribution capabilities. *1982 Pattern Recognition and Image Processing Conference* June, pp. 447–450.

Chapter Five

*Parallel Processing for Computer Vision**

Thomas A. Rice and Leah H. Jamieson

1. INTRODUCTION

Parallel processing has the potential of providing fast, flexible solutions to many computationally intensive tasks. In this paper, the use of parallelism for computer vision is described. Considerations for the design of a parallel architecture for computer vision are discussed.

The vision task consists of a number of different algorithms; several of the algorithms have markedly different computational characteristics. It is possible to achieve real-time implementations of some sequences of vision algorithms in hardware. The use of parallel processing allows significantly greater flexibility, both in the types of images that can be processed (e.g., gray-level images as well as binary) and in the choice of vision algorithms used. The work here presents theoretical analyses and simulation results for a collection of individual algorithms and for the overall vision task. This paper extends the work reported in Rice and Siegel [1].

2. DEFINITIONS FOR THE PARALLEL SIMULATION

In this section, two general models of parallel computation are defined, and the specific model used for the computer vision task is presented. The implementation of the parallel simulation is described.

* This research was supported by the United States Army Research Office, Department of the Army, under grant number DAAG29-82-K-0101.

INTEGRATED TECHNOLOGY FOR
PARALLEL IMAGE PROCESSING
ISBN 0-12-444820-8

2.1 Model

Single instruction stream–multiple data stream (SIMD) machines [2] represent a form of synchronous, highly parallel processing. Systems with up to 1 000 full processors have been proposed [3], [4]; systems with as many as 9 000 and 16 000 simple processors have been built [5], [6]. An SIMD machine typically consists of a *control unit*, a set of P *processing elements (PEs)*, each a processor with its own memory, and an *interconnection network*. The control unit broadcasts instructions to all PEs, and each active PE executes the instruction on the data in its own memory. The interconnection network allows data to be transferred among the PEs. SIMD machines are especially well-suited for exploiting the parallelism inherent in certain tasks performed on vectors and arrays.

Multiple instruction stream–multiple data stream (MIMD) machines [2] represent asynchronous parallel processing. MIMD systems with 16 [7] and 50 [8] processors have been built; MIMD systems with as many as 4 000 processors [9] have been proposed. An MIMD machine typically consists of P processors and M memories, M ≥ P, where each processor can follow an independent instruction stream. As with SIMD machines, there is a multiple data stream and an interconnection network. Thus, there are P independent processors that can communicate among themselves. There may be a coordinator unit to oversee the activities of the processors.

The parallel machine model assumed for the computer vision task consists of a set of PEs under the management of a control unit. The number of PEs is a power of two: $N = 2^n$. Each of the PEs has a unique address between 0 and $N - 1$. In addition, there exists an interconnection network to allow the simultaneous transfer of data among the PEs. For the computer vision task, the transfer patterns required will be uniform modulo shifts and cube interconnection functions. In a *uniform modulo shift*, PE j transfers data to PE $(j + d)$ modulo N for all j, $0 \leq j < N$, given a positive or negative integer distance d. The value of d may vary from one transfer to the next; however, for a given transfer, all PEs will send their data the same distance d. The set of *cube interconnection functions* consists of $n = \log_2 N$ functions, cube_i, for $0 \leq i < n$ [10]. If $P_{n-1} \cdots P_i \cdots P_0$ is the binary representation of a PE's address, then the cube_i function exchanges data between all pairs of PEs whose addresses differ in bit i:

$$\text{cube}_i(P_{n-1} \cdots P_i \cdots P_0) = P_{n-1} \cdots \bar{P}_i \cdots P_0$$

The model assumed here combines SIMD and MIMD attributes. Each PE contains the same code but executes the code on a different subimage. However, within each PE, the code can run in MIMD mode. This modification to the basic models allows faster execution on some code than a pure SIMD model, without incurring the expense of the full flexibility of an MIMD machine. The gains in speed will occur on the execution of conditional statements:

where < condition > do < block 1 >
elsewhere do < block 2 >

In SIMD mode, those PEs satisfying the < condition > execute < block 1 >. Then the remaining PEs execute < block 2 >. In the model here, < block 1 > and < block 2 > will be executed concurrently but in different sets of PEs. On the other hand, this is not full MIMD mode, as it is required that the code in each PE be the same. This aids in enforcing data coherence, e.g., insuring that a PE acquires the correct version of a variable from another PE.

Synchronization can take place in one of two ways. First, synchronization is required at all data transfer points, because data transfers often involve the same variable for all of the PEs. Even if the separate processors take different times to execute their code, they will be forced to synchronize at transfers to insure coherence. Explicit synchronization is also possible by one of the simulation language constructs that requires that all PEs finish a section of code before any can move to the next section of code.

The motivation for the assumed model comes from two directions. First, for many image processing operations, it is natural to consider executing the same code on subimages of the original image. Each subimage is a valid image, and the same types of operations are needed on the pixels of each subimage. Second, since the actual quantities of the various operations that will be performed on each subimage may vary, asynchronous operation may allow higher PE utilization.

This hybrid mode of operation may not be suitable for some algorithms. The requirements for such a mode to be useful are (1) that the PEs contain and execute the same code, with possible differences based only on the evaluation of conditional statements, and (2) that the need to synchronize at data transfers does not cancel the gains obtained by simultaneous evaluation of conditionals. For the vision algorithms examined here, these requirements are met.

2.2 Simulation

There are two major approaches to the development of parallel software. Either the software can be of a generally descriptive nature to illustrate the parallelism (or lack thereof) inherent in a task, or the software can be designed to be compilable and testable, either by parallel execution or serial simulation. Due to the computational intensity and intricacy of the computer vision task, the most reliable way to insure correctness is by testing. This guarantees that typical problem cases are being handled correctly by testing the software for a variety of images. A set of test images, some with multiple objects, was used for debugging and for analyzing computational speedup. Therefore, the software was designed so that it could be compiled and tested.

Programming was done in a modified version of the C language [11]. This language was chosen for the capabilities it provides for developing parallel data structures and the high degree to which one can manipulate system information (such as memory areas). The latter played a large part in the simulating of parallel data transfers. The actual conversion of the serial C language to a parallel language was done by means of macros and support subroutines. These features were designed to facilitate the development of parallel code without requiring the user to know the specific details of the serial implementation. Thus, one can simply use the macro file without knowing its details and can then write parallel code.

The major points of this implementation are as follows. A construct of the form

```
in_pe { codeblock; }
```

executes the enclosed block of code in each of the PEs. The prefix "PE." prepended to a variable indicates that the variable is local to a PE. All other variables are assumed to be global (i.e., the control unit has one copy of the variable). Global variables are used for such operations as loop control and overall conditional testing. There are also versions of the "in_pe" construct that allow the code to be executed in a limited subset of the PEs. These schemes use an address mask [12], which is a matching format that the PE address must match for execution to occur in that PE.

Interprocessor communication is accomplished by a *transfer* subroutine:

```
transfer (destination_address, source_address, offset)
```

The transfer routine uses these addresses along with information about the size and structure of the PE data space to simulate the transfer by a memory-to-memory move. Recursive transfers and broadcasts (in which one value is transferred to all of the PEs) are similar. Synchronization is needed at transfer points to insure data coherence.

The vision software and simulations were run on a dual-processor Vax 11/780 [13].

3. OVERVIEW OF THE VISION ALGORITHMS

In this section, an overview of the computer vision algorithms is provided. The parameters described are based on the SRI vision module [14] and Fourier descriptors [15].

A simple mechanism for entering an image into the system was desired. In the method chosen, the user employs a terminal with cursor control to draw an image on the screen and enter that image into the data memory. This section of the code used a small subsection of the "curses" [16] utilities available on the test system. This was later expanded to allow other image formats to be input. The images used here and in the subsequent steps are assumed to be binary images, although the algorithms can be generalized to handle gray-level images.

After an image has been entered into the data memory, the first task is to classify the image. This consists of transforming an image comprised of edge and non-edge pixels into an image with edge, internal, and external pixels. An internal pixel is a pixel that represents a point on an object, whereas an external pixel represents a point external to an object (such as the external background or a hole in the object).

After the inside and the outside of the image have been identified by the classification step, the holes in the image are located. A hole is defined as an area outside the object. Thus, the background also fits the definition of a hole. These holes are identified so that later merging can be accomplished easily. This capability is needed because holes that are initially thought to be separate may actually be joined.

The areas of the holes are computed and recorded at the same time as the original hole identification, because the data search patterns are similar. For purposes of isolating the object parameters, the background is defined to have an area of zero.

Once the inside of the object is known, the center of mass of the object is determined. Although in and of itself the center of mass is not a particularly useful parameter, it is used to normalize some of the perimeter statistics to be derived later.

To find the perimeter, the edge points that are adjacent to the background are identified. Once this has been done, it is a simple matter to find the distances from the perimeter points to the center of mass. These distances are used to calculate the average, minimum, and maximum perimeter distance from the center of mass.

Finally, using the already determined perimeter, a description of this perimeter is produced in the form of a list of coordinate pairs. This list can then be used to determine Fourier descriptors or other similar parameters.

Provisions have been made for the processing of images that contain multiple (nonoverlapping) objects.

4. DETAILED DESCRIPTION OF THE PARALLEL SOFTWARE

In this section, details of the vision algorithms and of their parallel implementation are presented. Results of the simulation of the parallel algorithms and analysis of the perfomance of the parallel vision system are presented in Section 5.

4.1 Image initialization

To be able to test the system easily, a simple method by which a user could enter an image into the system was developed. The user executes the vision program and then uses a standard keyboard to direct the cursor and draw an image border. The user also has the option of turning the cursor on and off to allow the drawing of unconnected borders (such as an internal border). The connection pattern for the drawing is an eight-neighbor scheme. That is, from a given point, the user can direct the cursor in any of the four horizontal and vertical directions as well as along the diagonals between these directions.

The screen size does not limit the size of the image being created, as the screen merely acts as a window into the image. During image creation the current position of the cursor is maintained in the upper left-hand corner of the screen. Messages and inputs are handled on the lowest line of the screen. If the drawing gets too near to any of the borders, the window into the image is automatically moved. The user can also specify a location to which to move the cursor. If this position is not in the current window, the window is automatically moved. All borders are strictly enforced: The user cannot draw beyond the edge of the border under any condition. After the user has created the image, an exit command automatically starts the image processing on the image.

In addition, images with 256 gray levels that are stored as character arrays (e.g., one character per pixel) can be loaded by the system. Simple thresholding routines as well as a Sobel operator are automatically applied to such images to convert them into binary images. The user is prompted for the thresholds for each image.

The produced image can be saved for later testing and can be reloaded and modified. The user also has the option of saving the results in a text file or of viewing the results as they are produced.

For the parallel implementation, once the image has been created, it is divided among the PEs with each of the PEs having an equally dimensioned stripe (either horizontal or vertical) of the image. Subsequently, each PE operates on the section of the image contained in its local memory, communicating with other PEs when further information is needed.

4.2 Internal/external classification

The internal/external classification labels each pixel as being on the inside of the object, outside the object, or on the border. The classification scheme implemented is a two-pass method. The first pass traverses the image from the upper left to the lower right. The initial classification of a pixel is based on the two neighboring points (to the left of the current point and above the current point) that have already been classified. The method tries to classify the new point as external if either of the previous points is external. If the adjacent points are both edges (border pixels), then information about the length of the edge and the previous region classifications are used to make the classification.

The second pass traverses the image from the lower right to the upper left (backward, as compared with the forward pass). This pass uses the four major compass points in relation to the current point to attempt to correct any classification errors. Again, the bias is toward external classification.

This section of the vision software uses several schemes to insure robustness. Besides the ability to reclassify points on the second pass, the software also looks for the specific case of tracing an edge. In addition, several trouble patterns are checked to prevent major misclassifications. Figure 1 illustrates the classification procedure. Figure 1(a) is the image before classification (border only). The edges are represented by '2.' Figures 1(b) and 1(c) are the image after the first and second passes of the classification, respectively. Internal points are represented by '1,' and external points are represented by '0.' An example of a reclassification on the second pass is illustrated by the outlined areas in Fig. 1(b) and 1(c).

In the parallel implementation, each PE works with its own stripe of the image data. The communication between PEs is limited to the values of the border elements of a subimage. One such transfer takes place for each border element on one of the sides of the subimage. These transfers are uniform modulo shifts of distance one. As the results show later, this section of the software demonstrates good speedup. Thus, the assumption of a two-pass classifier gives a conservative speedup estimation: if more passes were used, each pass would exhibit the same good speedup.

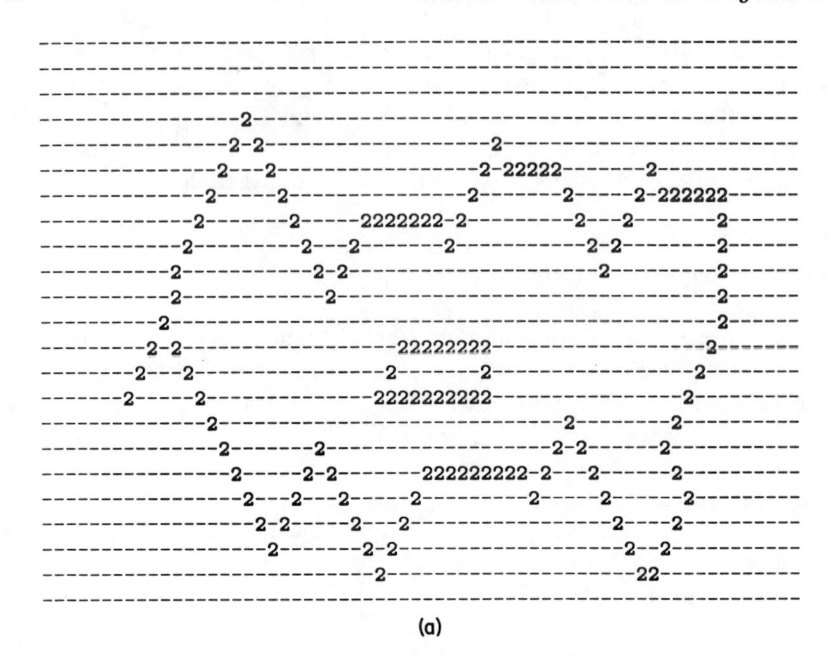

(a)

Fig. 1(a) Initial image

```
0000000000000000000000000000000000000000000000000000000000000000
0000000000000000000000000000000000000000000000000000000000000000
0000000000000000000000000000000000000000000000000000000000000000
0000000000000000020000000000000000000000000000000000000000000000
0000000000000000212000000000000000000020000000000000000000000000
0000000000000002111200000000000000000212222220000000200000000000
0000000000000021111120000000000000002111111120000021222222000000
0000000000000211111112000002222222021111111112000211111112000000
0000000000002111111111200021111111211111111111202111111112000000
0000000000021111111111120211111111111111111111121111111112000000
0000000000021111111111112111111111111111111111111111111112000000
0000000000211111111111111111111111111111111111111111111112000000
0000000002121111111111111111122222222111111111111111111120000000
0000000021112111111111111111200000002111111111111111111200000000
0000000211111211111111111112222222222111111111111111112000000000
0000000000000021111111111111111111111111111121111111120000000000
0000000000000002111111121111111111111111111202111111200000000000
0000000000000000211111202111111122222222212000211111120000000000
0000000000000000021112000211111200000000020000021111112000000000
0000000000000000002120000021112000000000000000002111122000000000
0000000000000000000200000002120000000000000000002112020000000000
0000000000000000000000000000200000000000000000000002200000000000
0000000000000000000000000000000000000000000000000000000000000000
```

(b)

Fig. 1(b) Classification: Pass 1

```
0000000000000000000000000000000000000000000000000000000000000000
0000000000000000000000000000000000000000000000000000000000000000
0000000000000000000000000000000000000000000000000000000000000000
0000000000000000020000000000000000000000000000000000000000000000
0000000000000000212000000000000000000020000000000000000000000000
0000000000000002111200000000000000000212222200000002000000000000
0000000000000021111120000000000000002111111120000021222222000000
0000000000000211111112000002222222021111111112000211111112000000
0000000000002111111111200021111111211111111111202111111112000000
0000000000021111111111202111111111111111111111121111111112000000
0000000000021111111111112111111111111111111111111111111112000000
0000000000211111111111111111111111111111111111111111111112000000
0000000002021111111111111111112222222211111111111111111120000000
0000000020002111111111111111120000000211111111111111112000000000
0000000200000211111111111111222222222211111111111111112000000000
0000000000000021111111111111111111111111111112111111112000000000
0000000000000002111111121111111111111111111202111111200000000000
0000000000000000211111202111111122222222212000211111120000000000
0000000000000000021112000211111200000000020000021111120000000000
0000000000000000002120000211120000000000000000021111220000000000
0000000000000000000200000002120000000000000000021120200000000000
0000000000000000000000000000200000000000000000000002200000000000
0000000000000000000000000000000000000000000000000000000000000000
```

(c)

Fig. 1(c) Classification: Pass 2

4.3 Identifying image holes

After the object has been separated from its surroundings by the classification operation, the holes in the image are identified. This process consists of two steps: initial local hole identification within each PE, followed by merging of holes between PEs. Initial hole labeling is first performed separately within each PE. This is done by creating a template array in each PE that is of the same size as the subimage in the PE. Each template location contains an identifier that indicates the local hole number for the corresponding subimage point, or zero for non-hole points. Each time an external point is located that is not adjacent to a previous hole, a new hole identifier is used and entered for that point in the template. If the external point is adjacent to a previous hole, then the previous identifier is continued. A two-neighbor scheme is used for all the pixels except those on one of the subimage borders. Since the points on one edge have only points from the previous row (or column, in the case of horizontal stripes) upon which to base a decision, a one-neighbor scheme is used at the borders. The software maintains a set of parameters that keeps track of merged holes and their statistics in order to handle the special case of an external point adjacent to two different previous hold identifiers. Experimentation showed that no accuracy problems were introduced by the small number of neighbors used in the classification.

These operations are performed totally within a PE: no communication with other PEs is needed. Each PE owns the information about its own holes. This information is transferred to other PEs during hole merging (described later). Figure 2 shows the internal hole identifiers for each PE. Hole identifiers that are adjacent (e.g., labels 3, 4, 5, and 6 in PE 2) are considered common. That is, only one of the identifiers contains the information for the hole. All of the others contain a pointer to the "master" information.

Once the holes have been identified in each PE, they are merged across the PE borders. This is done by transferring the borders of the PE hole template to adjacent processors and searching for matching holes. The areas are merged at the same time that holes are joined. In the scheme used, if a hole has only one edge on a PE border, then the statistics for that hole are transferred to that adjacent PE. This results in each hole being "controlled" by one PE. The information that needs to be transferred from each PE is placed on a transfer stack. These stacks are then transferred. All of these are transfers to logically neighboring PEs (uniform modulo shifts of a distance of one). The amount of information transferred is highly dependent on the

```
      PE 0             PE 1             PE 2             PE 3
1111111111111111 1111111111111111 1111111111111111 1111111111111111
1111111111111111 1111111111111111 1111111111111111 1111111111111111
1111111111111111 1111111111111111 1111111111111111 1111111111111111
1111111111111111 1-11111111111111 1111111111111111 1111111111111111
1111111111111111 ---1111111111111 111111-111111111 1111111111111111
111111111111111- ----111111111111 11111-------1111 111-111111111111
11111111111111-- -----11111111111 1111---------111 11--------111111
1111111111111--- ------11111----- --1-----------11 1---------111111
111111111111---- -------111------ ---------------1 ----------111111
11111111111----- --------1------- ---------------- ----------111111
11111111111----- ---------------- ---------------- ----------111111
1111111111------ ---------------- ---------------- ----------111111
111111111-2----- ---------------- ---------------- ---------3333331
11111111-333---- --------------33 22222----------- --------44444441
1111111-44444--- ---------------- ---------------- -------555555551
11111111111111-- ---------------- ---------------- ------6666666661
111111111111111- ---------------- ------------4--- -----77777777771
1111111111111111 -------4-------- -----------555-- ------7777777771
1111111111111111 2-----555------- 333333333-66666  -------777777771
1111111111111111 22---66666-----7 3333333333333333 -------777777771
1111111111111111 222-8888888---97 3333333333333333 2----8-777777771
1111111111111111 222222222222-::7 3333333333333333 22--999999999991
1111111111111111 2222222222222227 3333333333333333 2222222222222221
```

Fig. 2 Image hole determination

actual image. For purposes of easy identification and for separation of holes within an object from the background, the border background is defined as having an area of zero. The process of merging is illustrated in Fig. 3.

This method of merging holes across PEs is deterministic in that the maximum number of passes needed can be determined by the types of images being examined. For example, the more an object tends to spiral (a spring, for example, as compared with a wheel), the more passes are needed. To

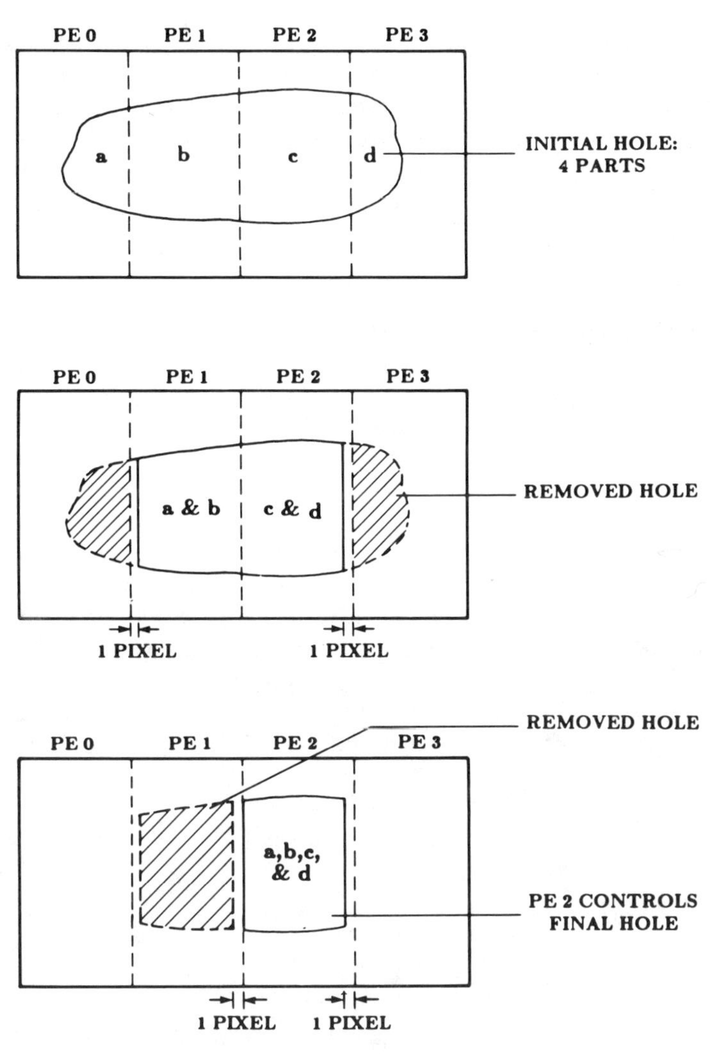

Fig. 3 Hole merging example

analyze performance, preliminary tests assumed a fixed number of passes (more than necessary for the images considered). In simulation, it was found that this section provides poor speedup. Thus, for this step, the net result of the fixed large number of passes is again a conservative estimate of the computational speedup of the algorithm. A refinement of the algorithm was also tested. By using only the required number of passes, appreciable improvements in speedup were obtained.

4.4 Computing image hole areas

The areas to be computed are tabulated at the same time as the hole identifiers are placed in the template in each PE. The area computation is therefore divided among the PEs. To handle the merging of holes, either within a PE or between PEs, an indirection table that points to the actual hole area is used.

4.5 Locating the center of mass

After the points that comprise an object are known, the center of mass of the object can be easily determined. In this system this step is performed by computing the moments in each PE separately and then summing across PEs using recursive doubling [17] (Fig. 4). The transfers used are the cube_i functions, $0 < i < \log_2 N$. This scheme requires that each PE know its absolute position in the configuration because the weighting of one of the moments in each PE is dependent upon the PE address. For example, if the stripes are in the vertical direction, then the x axis is split among the PEs. Moments that involve the absolute distance along the x axis depend on the PE address. To obtain the center of mass, $\log_2 N$ sets of transfers are needed. After the center of mass has been determined, it is broadcast to all PEs, because this information is needed at a local PE level in later processing.

4.6 Perimeter identification and perimeter statistics determination

Identifying the perimeter is straightforward once the external background hole has been identified. This hole has area zero by definition. An edge point next to an external hole (or next to another perimeter point) is a perimeter point. Since the area of holes is determined through an indirection table, all one needs to do is see if the hole has zero area. When a perimeter point is

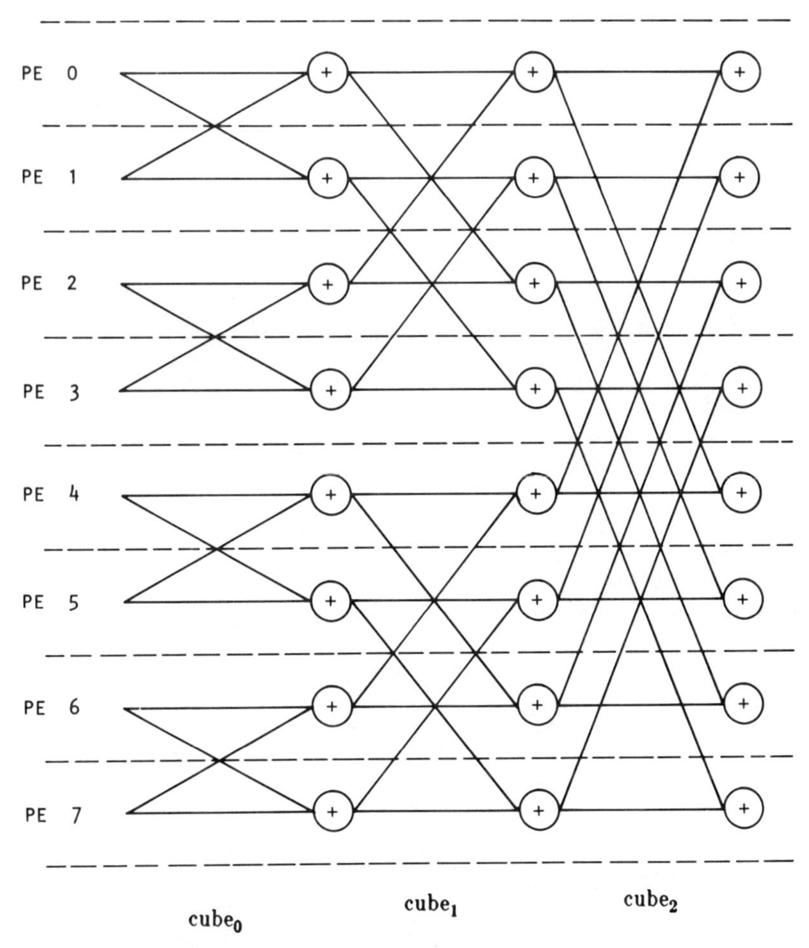

Fig. 4 Example of summing across PEs using recursive doubling

located in a PE, a counter in that PE is also incremented so that the total perimeter can be determined by a simple application of recursive doubling to accumulate the total across the PEs.

After the perimeter has been identified, it is a simple matter to find the distances between the perimeter points and the previously determined center of mass. This is done by scanning through the image template looking for perimeter points. Each PE scans its stripe of the image. For each perimeter point found, the radial distance from the perimeter point to the center of mass is determined. A running sum is kept of these distances, along with the minimum and the maximum distances. When the entire image has been

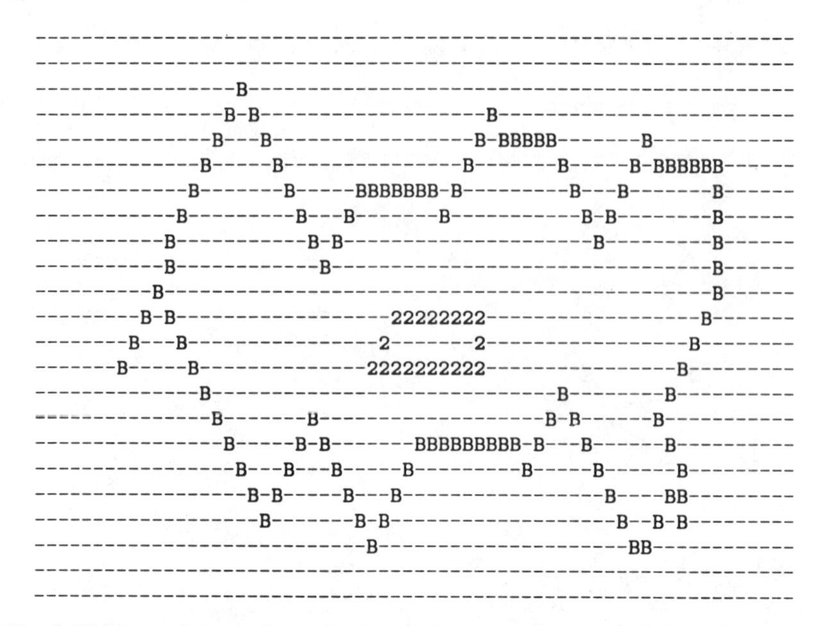

Fig. 5 Object perimeter determination and center-of-mass statistics. Total object perimeter is 109; the centre of mass, (33, 11); distances from center of mass to perimeter, min 3, max 24, average 12

```
000000000000000000000000000000000000000000000000000000000000
000000000000000000000000000000000000000000000000000000000000
000000000000000000000000000000000000000000000000000000000000
000000000000000020000000000000000000000000000000000000000000
000000000000000212000000000000000000200000000000000000000000
000000000000002111200000000000000000000212222200000020000000000
000000000000021111120000000000002111111120000021222222000000
000000000000211111112000022222220211111111200021111111112000000
000000000002111111111200021111121111111111202111111112000000
000000000021111111111120211111111111111111121111111112000000
000000000021111111111112111111111111111111111111111112000000
000000000211111111111111111111111111111111111111111112000000
000000002021111111111111112222222211111111111111111120000000
000000020002111111111111111200000021111111111111111200000000
000000200000211111111111112222222222111111111111112000000000
000000000000021111111111111111111111111111211111112000000000
000000000000002111111211111111111111111120211111200000000000
000000000000000211112021111112222222221200021111120000000000
000000000000000021112002111112000000002000002111112000000000
000000000000000002120000211120000000000000002111122000000000
000000000000000000200000021200000000000000000211202000000000
000000000000000000000000002000000000000000000022000000000000
000000000000000000000000000000000000000000000000000000000000
```

Fig. 6 Example of vision software output. Two holes in image; total perimeter is 109; total hole area 7; center of mass (33, 11); distances from center of mass to perimeter, min 3, max 24, average 12; one object in image

scanned, recursive doubling is used to find the average, minimum, and maximum distances. Three stages of recursive doubling transfers are needed, one set for each of the perimeter statistics being gathered. This results in a total of $3 \log_2 N$ transfers.

Figure 5 shows the identified perimeter for an image. The perimeter (border) is noted by "B," as compared with "2" for a nonperimeter edge point. Figure 6 shows an example of the overall output of the vision software.

4.7 Data preparation for Fourier descriptors

As an illustration of some of the higher level functions that can be performed once the basic parameters have been extracted, the image can be converted into the information necessary to calculate Fourier descriptors [15]. This information is simply an ordered list representation of the perimeter of the object. Each entry in this list consists of a set of coordinates representing a perimeter point. Fourier descriptors have been proposed as a method of performing shape analysis.

The vision software begins this step by forming the perimeter nodes into a multiply linked list, which facilitates the removal of false perimeter points (spikes). This converts the perimeter into a traceable contour. Next, these linked lists are transferred to one PE which completes the processing. This requires uniform modulo shifts of distances from 1 to $N - 1$. This processing includes converting the lists into partial ordered lists and then combining these lists. Other schemes, such as forming the partial lists in each PE separately, were found to induce such a large amount of overhead in transfers that any advantages in parallelism were lost. The final contours in the single PE are then broadcast to the remainder of the PEs in preparation for the Fourier descriptor calculations. If the perimeter is equally distributed among the PEs, $(N - 1)/N$ of the partial ordered listings need to be transferred. Each of the objects in one of these lists contains ten data fields (two link fields for the linked list and eight neighbor pointers) If the perimeter is not equally distributed, then the perimeter could be gathered into the PE with the largest number of perimeter points, and this requires fewer total transfers. Thus, if there are P perimeter points, a maximum of $(N - 1) P/N$ transfers are needed.

4.8 Multiple object images

The software that has been described has treated the content of the image field as one object. If there is more than one (nonoverlapping) object in the image field, the same software can still be used, but the results will be a composite of the information for the separate objects. However, it is not exceedingly difficult to separate the information for the separate objects.

Once the contours of the image have been determined, the software knows how many separate objects are in the image. This involves the classification, hole and area identification and merging, and perimeter determination steps described above. That is, the number of contours equals the number of objects in the image, given that the objects do not overlap and that no object is inside another (such as a bolt in a wheel rim). The items can be processed individually by removing the objects corresponding to the undesired contours and reprocessing the image. This can be done for each object in the image. The individual processing involves all the the previous steps, from classification through perimeter determination and perimeter statistics.

To remove an object from the image, its perimeter points (known from the contour) are marked to be removed. Two passes are made over the image (similar to the initial classification) to convert internal, perimeter, and edge points bordering the removal points to removal points themselves. This is similar to the erosion scheme used by CLIP4 [18]. A final pass is made over the image to convert all removal points to external points, effectively erasing the object from the image.

If the program detects multiple images, it still gives the composite results, but it also sequentially erases all but one of the objects and then processes the remaining object. This additional processing is identical to the main processing sequence, except that the checks for multiple objects are omitted.

4.9 Additional parameters

Other parameters can be added to a vision system to improve the robustness of object identification. Some of these additional parameters are simply combinations of previous parameters. An example of such a parameter is the factor of roundness (how circular the image is), which is computed by dividing 4π times the area by the square of the perimeter. The area of the object could also be calculated at the same time that the second pass in the internal/external classification step is made. This area could be combined with the internal hole area to provide a total of the areas occupied by the object. The ratio of hole area to total area is similarly obtainable.

There are other parameters that would require additional computation in the main processing sequence. This class of parameters includes such features as second moments, ratios of major and minor axes, finding the bounding rectangle, and line fitting. Others could be added, based upon the specific task at hand.

Finally, one needs to consider the nonideal cases in which multiple objects in the image overlap or the objects are not entirely contained within the borders. Much information for the latter case can be obtained from processing the object as usual and then applying statistical methods to determine possible matches with known objects. The other case is not as simple: Some type of image reduction is necessary if it is determined that an object is not known. Such software could selectively reduce protrusions of an object until a known object is identified.

5. ANALYSIS

To evaluate the use of the parallel architecture for computer vision, analytical comparisons of the parallel and serial algorithms were performed, and the simulation of the parallel software was compared to the serial implementation. An estimation of the computational speedups was derived by an examination of the structure of the parallel software. Table I summarizes the speedups for the major algorithms. The proportions of time required by different sections of the code were determined by executing a serial version of the algorithm. (The serial algorithm does not incur any overhead for operations such as transfers or processor disabling.) The time proportions are used to provide a weighting of the parallel speedup results. In this way, a section with low speedup that requires only a small fraction of the serial processing time does not falsely lower the overall speedup. Similarly, a section with high speedup that requires only a small fraction of the serial processing time does not falsely raise the overall speedup. With the time

Table 1
Computational performance results

Algorithm division	*Approx. speedup*	*Serial time*	*Time proportions*
Class()	$N(I/(I+N-1))$	15.36	0.3531
Holes()	$N/((N-1)(\text{SPIFAC}+1))$	15.79	0.3630
Areas()	(called by holes)	N/A	N/A
Center()	N	1.64	0.0377
Pstats()	N	10.71	0.2462

Note: I = Image border ($I \times I$ image); N = Number of PEs; SPIFAC = number of times a section of the object in the image can switch directions in crossing the image (for example, the letter Z would have a SPIFAC of 2). For the images analyzed, SPIFAC = 6.

Table 2
Parallel simulation: experimental results (64 × 64 image)[a]

Algorithm	*1 PE*	*2 PEs*		*4 PEs*		*8 PEs*	
	Avg	*Avg*	*Norm*	*Avg*	*Norm*	*Avg*	*Norm*
Class	40.25	42.25	21.13	50.25	12.63	58.67	7.333
Holes and areas	48.75	79.75	39.88	183.0	45.75	642.0	80.25
Center	5.25	6.25	3.125	6.5	1.625	7.0	0.875
Pstats	14.75	15.25	7.625	12.75	3.188	15.33	1.917
Time subtotal	109.0		71.76		63.19		90.38
Partial speedup	1		1.52		1.72		1.21
Chain (serial)	64.75	23.75	23.75	28	28	27.67	27.67
Chain (parallel)	N/A	76.75	38.38	139.25	34.81	272.33	34.04
Total time	173.75		133.89		126.0		152.09
Overall speedup	1		1.30		1.38		1.14
Efficiency	1		0.65		0.345		0.143

[a] *Times in 1/60th second*

proportions, the total weighted speedup $S(N)$ for processing an $I \times I$ image using N PEs can be computed:

$$S(N) = \frac{0.3531NI}{I + N - 1} + \frac{0.3630N}{(N - 1)(SPIFAC + 1)} + 0.0377N + 0.2462N$$

$$= N\left[\frac{0.3531I}{I + N - 1} + \frac{0.3630}{(N - 1)(SPIFAC + 1)} + 0.2839\right]$$

The experimental results for the major sections of the software are presented in Table 2. The columns labeled Avg give the average time the serial simulation took for each step of the algorithm. The columns labeled Norm give the conversions of the average serial times to the average parallel times. This is the normalized execution time. The Time-subtotal row indicates how much time the first four component algorithms (internal/external classification, hole identification assuming a fixed number of passes, center of mass and perimeter statistics) required. The speedup that these partial times indicate is presented in the Partial-speedup row. The final algorithm step, formation of the chain code representation of the perimeter, is represented by two rows in the tables, because it has both a serial and a parallel component. Finally, the Total-time and Overall-speedup rows indicate the time that the entire processing operation needed and the speedup reflected by this time.

One additional measure of the performance of a parallel algorithm is the *efficiency* $E(N)$, defined to be the ratio of the speedup to the number of processors [19]. Table 2 also shows the speedup for the case of a 64 × 64 image. For the example, although the speedup increases with N, the rate of increase is not proportional to N, and the efficiency decreases fairly sharply with N.

The simulations were designed to provide a conservative estimate of the speedup; assumptions about transfer timings and synchronization delays were approximated. The problem of nondeterminism in speedups was handled by using deterministic versions of nondeterministic routines. Again, these routines were designed to provide a conservative estimate of the speedup. No overlap of processing and transfers was assumed, although in many situations, inter-PE transfers can be performed at the same time that independent processing is occurring. The simulation results can, therefore, be used as a rough indicator of the speedup obtained by the parallel algorithms. Both the analytic and experimental results bear out the observation that the speedup will not grow as N, because the algorithms in which the largest proportion of time is spent (hole merging and chain code formation) have less than ideal speedup. (The experimental speedups are somewhat less than the analytic speedups due to the conservative assumptions made throughout the simulation.) In particular, the discrepancy between the theoretical and the experimental results is primarily in the holes and areas section. In this section, the theoretical results take into account the number of times the merging must be performed but do not take into account the overhead incurred by the transfers required by the merging. This overhead turns out to be a substantial portion of the algorithm, to the extent that it destroys the effectiveness of the increased parallelism. It appears that having subimages less than 16 pixels wide is counterproductive.

To address the problems with the hole merging algorithm, a new version of this algorithm was constructed that performs only the required number of hole merging steps (thus removing one of the earlier conservative assumptions). The algorithm is divided into two parts, which correspond to single-sided hole merging (such as was illustrated earlier) and multiple-edged hole merging (which handles ringlike holes such as the background hole). Each of these stages proceeds until the number of holes merged in each PE is zero. This has the advantage of eliminating unneeded overhead as well as having the capability of dealing with pathological cases that might require additional merging steps.

The results for this software with this modification included are in Table 3. Note that with this modification, for 64 × 64 images, eight processors still provide speedup gains, whereas previously only two or four could be used before the results deteriorated due to the overhead of the parallelism. With eight processors, the stripes in each PE are only eight pixels wide, so the proportion of time spent in overhead to coordinate between PEs is

Table 3
Non-deterministic merging parallel simulation results (64 × 64 image)[a]

Algorithm	*1 PE* *Avg*	*2 PEs* *Avg*	*Norm*	*4 PEs* *Avg*	*Norm*	*8 PEs* *Avg*	*Norm*
Class	40.25	45.25	22.63	49.0	12.25	58.67	7.333
Holes and areas	48.75	83.0	41.5	121.25	30.31	249.0	31.13
Center	5.25	6.5	3.25	6.75	1.688	6.67	0.8334
Pstats	14.75	15.75	7.875	14.0	3.5	15.33	1.917
Time subtotal	109.0		75.25		47.75		41.21
Partial speedup	1		1.45		2.28		2.64
Chain (serial)	64.75	24.0	24.0	28.0	28.0	28.33	28.33
Chain (parallel)	N/A	81.0	40.5	144.25	36.06	272.67	34.08
Total time	173.75		139.76		111.81		103.62
Overall speedup	1		1.24		1.55		1.68
Efficiency	1		0.62		0.388		0.210

[a] *Times in 1/60th second*

substantial. Thus, simulation demonstrated that the major problem with the parallel implementation is basically of one form: The number of transfers needed reduces the effectiveness of the parallelism. This can occur when the amount of information that is needed to make a proper decision (such as for hole merging) is large. This problem can manifest itself in several forms, such as algorithms that are inherently serial or that require data from the entire image. Such tasks might better be performed in one PE or in the control unit.

6. ARCHITECTURAL CONSIDERATIONS

A specific type of architecture has been assumed throughout this simulation and analysis. At this point, this restriction will be removed, and the tasks considered will be examined to explore a parallel architecture tailored to the characteristics of the vision task.

By examining the algorithms, one can see that a given memory area (the memory to be accessed in an interleaved manner, further improving system processing section. If the memory is dual-ported, with one *write* channel and two *read* channels, then the need for transfers can be virtually eliminated. In such an approach, the memory that was previously the exclusive responsibility of a specific PE would still be connected to that PE by the *write* channel and one of the *read* channels. However, the other *read* channel would be connected to a memory redirection network that would be setable by the control unit when a new type of access pattern is needed. This redirection network could be either bidirectional or (more practical) two unidirectional networks, one direction being used to transmit the memory addresses and the other being used to return the data. The advantage of using two unidirectional networks is that information can be flowing in both directions at the same time without the need for redirection or buffering. This would allow the memory to be accessed in an interleaved manner, further improving system performance. When this scheme is compared with the number of transfers needed in some of the processing steps (such as in hole merging and Fourier descriptor preparation), the possible savings are evident.

7. SUMMARY

In this paper, analytic and simulation results for the application of parallel processing to the computer vision task have been presented. Because of the modular design of the software developed, it is possible to expand the processing sequence to include other common image processing techniques. From the analytic and simulation capabilities described, given specific speed requirements for a particular vision task and assumptions about processor speed, it will be possible to determine the number of processors needed to satisfy the task requirements. This work contributes to the understanding of the design of parallel systems for image processing applications.

REFERENCES

[1] Rice, T. A., and Siegel, L. J. (1983). Parallel algorithms for computer vision, *1983 Comp. Soc. Workshop on Computer Arch. for Pattern Analysis and Image Database Management*, **2**, October, pp. 93–100.

[2] Flynn, M. J. (1966). Very high-speed computing systems, *Proc. IEEE*, **54**, December, pp. 1901–1909.

[3] Pease, M. C. (1977). The indirect binary *n*-cube microprocessor array, *IEEE Trans. Comp.*, **C-26**, pp. 458–473.

[4] Siegel, H. J. *et al.*, PASM: A Partitionable SIMD/MIMD system for Image Processing and Pattern Recognition, *IEEE Trans. Comp.*, Vol **C-30**, December, pp. 934–947.

[5] Batcher, K. E. (1980). The design of a massively parallel processor, *IEEE Trans. Comp.*, **C-29**, September, pp. 836–844.

[6] Duff, M. J. B. (1982). Parallel algorithms and their influence on the specification of application problems, In *Multicomputers and Image Processing: Algorithms and Programs* (K. Preston and L. Uhr, eds). Academic Press, New York.

[7] Wulf, W., and Bell, C. (1972). C.mmp—A multi-miniprocessor, *AFIPS 1972 Fall Joint Comp. Conf.*, December, pp. 765–777.

[8] Swan, R. J., *et al.* (1977). The implementation of the Cm* multi-microprocessor, *AFIPS 1977 Nat'l. Comp. Conf.*, June, pp. 645–655.

[9] Gottlieb, A., *et al.*, (1983). The NYU ultracomputer—designing an MIMD shared memory parallel computer, *IEEE Trans. Comp.*, **C-32**, February, pp. 175–189.

[10] Siegel, H. J. (1984). *Interconnection Networks for Large Scale Parallel Processing: Theory and Case Studies*, D. C. Heath and Co., Lexington, Massachusetts.

[11] Kernighan, B. W., and Ritchie, D. M. (1978). *The C Programming Language*, Prentice-Hall, Englewood Cliffs, New Jersey.

[12] Siegel, H. J. (1977). Analysis Techniques for SIMD Machine Interconnection Networks and the Effects of Processor Address Masks, *IEEE Trans. Comp.*, Vol. **C-26**, February, pp. 153–161.

[13] Goble, G. H., and Marsh, M. H. (1982). A Dual Processor Vax** 11/780, *IEEE 9th Annual Symp. on Comp. Arch.*, April, pp. 291–298.

[14] Nitzan, D., *et al.*, (1979). Machine intelligence research applied to industrial automation, SRI Report, Menlo Park, California, August.

[15] Wallace, T. P., and Mitchell, O. R. (1980). Analysis of three-dimensional movement using Fourier descriptors, *IEEE Trans. Pattern Analysis and Machine Intelligence*, **PAMI-2**, November, pp. 583–588.

[16] Arnold, K. Screen Updating and Cursor Movement Optimization, A Library Package, Unix* Version 4.1 bsd.

[17] Stone, H. S., ed. (1980). *Introduction to Computer Architecture*, Science Research Associates, Chicago, Illinois, pp. 394–396.

[18] Reynolds, D. E., and Otto, G. P. (1981). Software tools for CLIP4. Report No. 82/1, Dept. of Physics and Astronomy, University College, London, January.

[19] Kuck, D. J. (1977). A survey of parallel machine organization and programming, *Computing Survey*, **9**, March, pp. 29–59.

* Unix is a trademark of Bell Laboratories.

** VAX is a trademark of Digital Equipment Corporation.

Chapter Six

An Overview of Image Algebra and Related Architectures

Stanley R. Sternberg

1. INTRODUCTION

Recently a shift of direction has been observed in the design of commercial vision systems for robot control and industrial inspection. Several companies have announced or released systems that incorporate image or operational parallelism as their principal technique for image processing. Two-dimensional arrays of identical processing elements and pipelines of processing stages are now moving out of the laboratory and into industrial applications. The architectures of these systems are far removed from the pattern-recognition systems that have dominated industrial vision research for the past ten years or so. This paper outlines the progress made to date on the development of machine vision systems for image processing and analysis based on image algebraic programming and pipelined cellular logic architectures.

The paper first takes an historical perspective, examining the sequence of developments leading up to current designs, and indicates the motivations of the designers. Next, we examine a particular realization of a parallel image processor architecture and its programming language. Because the goal of all image processor designs is to replace or enhance human visual perception in critical or exhaustive applications, the paper concludes with a detailed description of a machine vision inspection application.

2. PARALLEL ARCHITECTURES FOR MACHINE VISION

Cellular logic image processor designs are heavily influenced by our perceived notions of information processing in natural systems. Even so, the earliest

INTEGRATED TECHNOLOGY FOR
PARALLEL IMAGE PROCESSING
ISBN 0-12-444820-8

studies of interative image transformations in cellular space by von Neumann [1], Ulam [2], and Unger [3] focused exclusively on the nonperceptual mechanics of universal computation and self-reproduction. Interest in image processing in cellular arrays did not flourish until Hubel and Weisel [4] established the physiological basis for iterative cellular transformations as models of human visual perception. Theoretical studies of parallel image processing in cellular automata were not undertaken until some time later by Rosenfeld [5]. Early cellular models of vision were simulated on general-purpose computers. Later, prototype hardware was constructed in the form of small two-dimensional arrays, for example, the SOLOMON 16 × 16 cellular array created by Slotnick [6]. An alternative to the two-dimensional array architectures was constructed by Serra and Klein, called the Texture Analyzer System (TAS) [7]. The TAS employed a pipeline organization of operators which were scanned over an image by suitably configured shift register delays. Preston [8] and Golay [9] also employed a pipeline approach in the design of the Diff3 blood-cell analyzer [10]. Motivated by applications requiring both real time and high resolution, Sternberg [11] developed a pipelined image processor architecture, the Cytocomputer [12], which extended many of the algorithmic features of the TAS and Cellscan machines into the domain of gray-scale processing [13]. Recently, Kung [14] has generalized the pipeline architectural concept, referring to it as a systolic array of processing elements.

The 2-D cellular logic approach to image processing is indicated by considerations of ultimate parallelism in the design of high-speed, low-cost identical components. This goal has been realized to a considerable extent in Batcher's Massively Parallel Processor (MPP) [15] and Duff's Cellular Logic Image Processor (CLIP) [16] cellular array machines.

Experience with these and other very high-speed image processors indicate that high speeds are only achieved while the image data is in the cellular array. A number of useful computations that one would like to perform involve processing that cannot be done in the cellular array. A significant fraction of the time is, therefore, spent in loading and unloading image data from the cellular array. It is in this image I/O interface that the basic problem of higher speed lies.

3. LANGUAGES FOR MACHINE VISION

The parallel architectures previously discussed are general purpose in that they rely on application software to configure them for a particular image processing task. To develop application software quickly and accurately, image processing languages are developed for each processor architecture, compiling user commands into machine executable code. Generally, no language developed for one machine will run on another. Still, there are underlying language structures that incorporate some general principles. One of these structures is called *image algebra*, or *mathematical morphology*.

Mathematical morphology is the name given to the science of image analysis that utilizes set theoretic descriptions of images and image transformations. The fundamentals of mathematical morphology were developed by Matheron [17] at the École des Mines in France and were based on the geometry of Minkowski [18]. Serra [19] conceived of the first image processor incorporating morphological image processing, the TAS. Because the morphological approach to vision systems offers inherent advantages of speed and flexibility, the method is rapidly gaining in popularity and acceptance.

As a means of analyzing texture, porosity, granularity, and other realizations of random sets, mathematical morphology was originally developed to determine the parameters of the distributions of these random sets. Mathematical morphology incorporates the two principal components of any computer vision system, image transformation and feature measurement. As a rule, it is not concerned with classification of measurements. Contrast this with the approach of many commercial systems developed in America that focus primarily on measurement and classification.

How does one develop the solution to a machine vision problem? More specifically, how does one develop software for a particular machine vision application? Most frequently, because of the lack of standardized methods and approaches to machine vision problems, the interactive method is employed. Working with camera, monitor, and development hardware and software, the programmer attempts to incorporate his own visualization of the algorithmic solution into the system. A framework of preexistent concepts and system constraints structures his approach to problem solution and creates specific channels in his interaction with development system hardware and software. Here, there arises a critical issue. To couple machine processing efficiently with the human programmer's visual system, there needs to be a close link between what the programmer sees and what he is able to generate readily in terms of image transformations. When one can clearly visualize the desired result but is unable to program the algorithmic solution, frustration is the product.

4. GENESIS 2000 MACHINE VISION DEVELOPMENT SYSTEM

Generally, pipelined architectures, like their two-dimensional array counterparts, consist of identical modules. The image-flow architecture, which is the basis of the Genesis 2000 machine vision development system, differs in that each of its elements are specialized for a different image processing function. Because all images processed in an image-flow architecture pass through the same pipeline of processing functions, the architecture superimposes a structure of iterative procedures on image processing software development.

In the general methodology of structured iterative procedures, the functions are

(1) Select and compare two images and extract their differences.

(2) Distinguish significant features in the difference image.

(3) Count, locate, and otherwise quantify the distinguished features as a set of numbers constituting the relevant data in an image.

(4) Control a real-world, real-time process and determine the next image processing procedure based on the extracted relevant data.

These four basic processing functions are implemented in four basic components of the image flow architecture. The components are

(*Arithmetic Logic Unit* (*ALU*). Add, subtract, multiply, divide, or logically compare two images or an image and a constant.

Geometric Logic Unit (*GLU*). Remove those pixels of an image that do not conform to a programmable shape and intensity specification. Single or multiple morphological transformations occur in a single GLU. Multiple GLUs can be inserted in the image flow for higher speed.

Count and Locate Unit (*CLU*). Count the number of occurrences of pixels in a specified rectangular window or intensity interval. Locate the coordinates of all pixels in a given state.

Central Processing Unit (*CPU*). Control the ALU, GLU, CLU, the other development system components such as digitizers and frame buffers, and peripheral devices such as robots, cameras, and material handling equipment. Control decisions are made according to programmed criteria and quantitative image descriptors extracted by the CLU.

A block diagram of the MVI Genesis 2000 hardware is shown in Fig. 1. The system consists of two subsystems, the Image Flow Processor and the Motorolo 68000 Host Microcomputer. The components of the Image Flow Processor are shown in the system block diagram connected by the Video Bus; the components of the 68000 Microprocessor subsystem are interconnected by the Intel Multibus.

The Image Flow Processor consists of an Analog/Digital Converter that provides real-time conversion of both the analog video signals and the digital video signals, and a recirculating pipelined image processor whose components are a digital Frame Buffer Memory, an Arithmetic Logic Unit (ALU), a Geometric Logic Unit (GLU), and a Count/Locate Unit (CLU). The Analog Digital Converter additionally includes both input and output lookup tables for effecting contrast enhancement, binary thresholding, pseudocolor display, and so forth. The pipeline data flow rate is 10 megabytes/second.

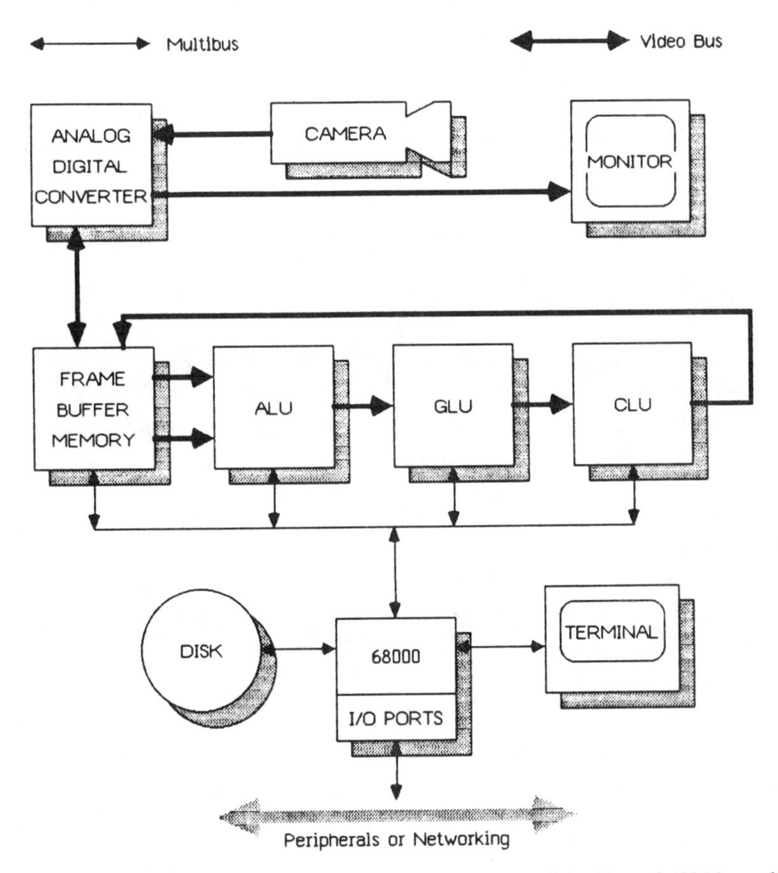

Fig. 1 Block diagram of the vision development system of the Genesis 2000 machine

The Frame Buffer Memory consists of three or more 512 × 512 frame buffers, denoted 0., 1., 2., etc., each frame buffer being eight bit-planes deep. Individual bit planes are denoted as 0.0, 1.7, etc., with the buffer designation followed by the bit-plane designation separated by a decimal point.

5. GLU: PIPELINED PARALLELISM

A unique element of the image flow architecture is its Geometric Logic Unit (GLU). The GLU implements the morphological operations on both gray-scale and binary images. It does so in an extremely efficient manner that minimizes both processing time and hardware complexity.

The principal transformations of the mathematical morphology are called dilations and erosions. They are most familiar in terms of the swelling or shrinking of binary blobs, although gray-scale dilations and erosions are also important in gray-scale image processing [22]. The terms *dilation* and *erosion* are most frequently applied to the isotropic expansion or contraction of a blob. In the notions of mathematical morphology, we refer to isotropic swelling or shrinking as dilation or erosion by a disk. The disk represents the dilation of an image consisting of a single point and is called a structuring element. The radius of the disk determines the degree of swelling or shrinking.

Dilation by other shaped structuring elements is also possible. In general, the extent of a structuring element in a given direction determines how far an image will be swelled in that direction. Because image dilations and erosions can require extensive computations, it is important that the algorithm for their implementation be as efficient as possible. To understand the alternatives and costs associated with a given algorithm, we must begin with some definitions.

Minkowski [18] first described dilation as a union of translations of an image X to all points b belonging to the structuring element B, referred to as Minkowski addition,

$$X \oplus B = \bigcup_{b \in B} X_b$$

where X_b denotes the translation of the set X to the point B. We see by this formulation that the dilation of an image X by a structuring element B generally involves the union of $N(B)$ translations of X, where $N(B)$ is the number of points in B. If one of the points in the structuring element B is the origin, then dilation according to the Minkowski addition formulation requires $N(B)$-1 distinct translations of X to be combined through the union operation with X. Likewise, if we take the form of the erosion of X by B to be that of Minkowski subtracting—namely,

$$X \ominus B = \bigcap_{b \in B} X_{-b}$$

the number of unions of translations to effect an erosion is generally equal to $N(B)$, the number being reduced to $N(B) - 1$ when the origin belongs to B.

Specialized pipelined computer architectures for computing image dilations and erosions are able to perform their specialty in considerably fewer than $N(B)$ iterations. This results from the algebraic morphological property called *structuring element chaining*. The chain rule for dilations and erosions states that if structuring element B can be formulated as the dilation of structuring elements B_1 and B_2, then dilating X by B is equivalent to interatively dilating X by B_1 and then B_2,

$$X \oplus B = [(X \oplus B_1) \oplus B_2]$$

where $B = B_1 \oplus B_2$. Similarly, for the case of erosion,

$$\mathrm{X} \ominus \mathrm{B} = [(X \ominus B_1) \ominus B_2]$$

where $B = B_1 \oplus B_2$.

As an example of the use of structuring element chaining, consider the implementation of hexagonal dilations and erosions by the TAS I, or Texture Analyzer System [7]. The TAS performs dilations by hexagonal structuring elements by chaining dilations or erosions of line segments. Figure 2 illustrates this concept for a regular hexagonal structuring element. The hexagon of Fig. 2 consists of 331 points, including the origin, requiring 330

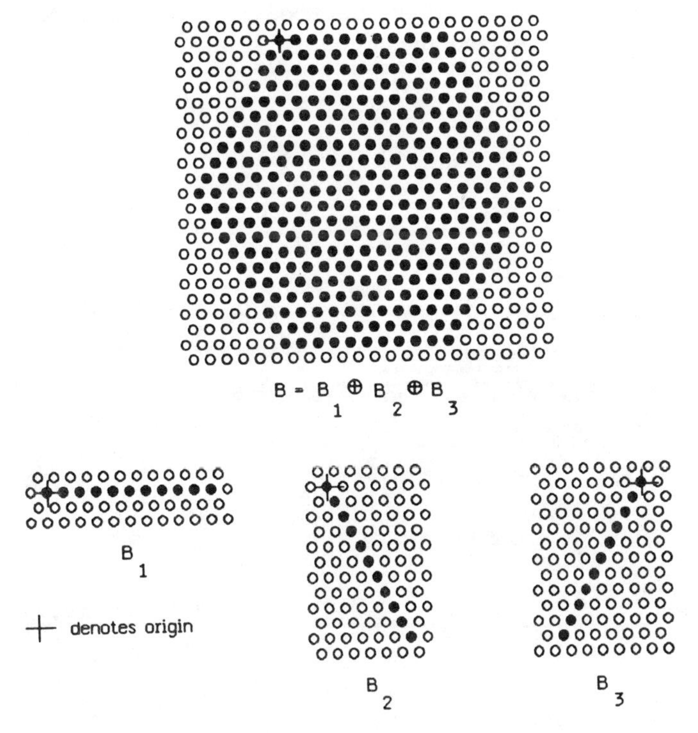

Fig. 2 Dilation by hexagonal structuring agent element B implemented as a sequence of three linear dilations, as performed in the Texture Analyzer System [7]

iterations of translation and union to implement a dilation or erosion by directly applying the Minkowski formulation.

The hexagonal dilation is performed in the TAS by a pipeline of three stages, each stage dilating the image by a line segment of programmable length in the three primary directions of the hexagonal grid. To accomplish the hexagonal dilation illustrated in Fig. 2 requires three successive linear dilations, each linear dilation consisting of 10 interactions of translation and union, for a total iterative step count of 30. This early implementation of the TAS only dilated and eroded by hexagons. Later versions performed neighborhood operations like the Cytocomputer below.

Another pipelined architecture, the Cytocomputer [12], also achieves a considerable savings of processing steps by chaining neighborhood operations. In a Cytocomputer pipeline, each stage of the chained dilation is an erosion or dilation by a neighborhood structuring element. Cytocomputer neighborhood structuring elements are subsets of either 3×3 square neighborhoods or 7-element hexagonal neighborhoods. Figure 3 illustrates the Cytocomputer dilation by a duodecagon structuring element implemented as a chain of eight neighborhood dilations. Each neighborhood structuring element contains the origin, and so it is seen that the resulting

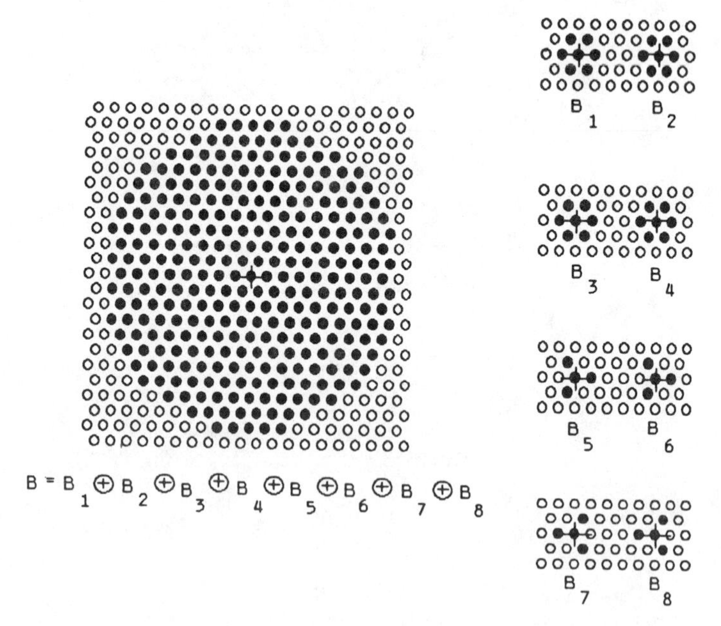

Fig. 3 Dilation by duodecagon structuring element B implemented as a sequence of eight neighborhood dilators, as performed by the Cytocomputer [12]

duodecagonal dilation is achieved by four neighborhood dilations of six translation and union steps and four neighborhood dilations of three translation and union steps for a total iteration count of 36.

The GLU chains structuring elements that consist of two points, one of these points being the origin, and the other being an arbitrary point in the digital grid. Figure 4 illustrates the sequence of chained dilations for producing the 24-sided structuring element shown. The total iterative count here is 15; each successive step in the chained dilation sequence in accomplished by a single translation and union of the prior result. This dilation

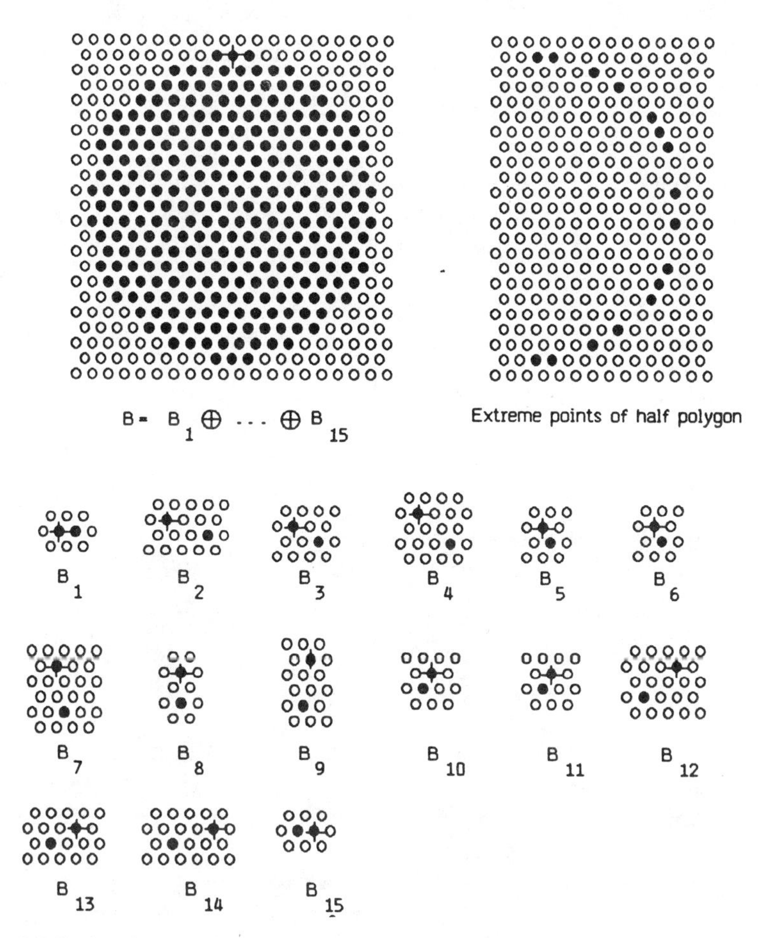

Fig. 4 Dilation by 24-sided polygonal structuring element B implemented as a sequence of 15 pairwise dilations, as performed by the Genesis 2000 [21]

operation as performed in a GLU is called image *cloning*, or simply cloning, because each step of the chained dilation sequence creates a duplicate of the previous result at some offset location before combining it through the union operation back into the result of the previous step. Similarly, a chained erosion in a GLU is referred to as an image *collapse*, or simply collapsing.

6. BLIX: BASIC LANGUAGE OF IMAGE X-FORMATION

The MVI Genesis 2000 differs from most other vision systems in that it is fully programmable in a high-level, integrated language called BLIX [21]. The BLIX language is implemented in the Genesis 2000 as a custom shell in the UNIX operating system. Thus, in addition to the image processing capabilities of BLIX, Genesis 2000 users also have available all the special features of the powerful UNIX System V operating system, including visual editing and hierarchical file directories.

BLIX commands are based on the operations of mathematical morphology, the basic method of image analysis based on set theory. The inspection algorithm illustrated in the following section employs the following BLIX commands:

and
or
lgt
implode
clone
open

In addition to commands that implement the morphological image transformations, the following BLIX utility commands are employed in the inspection algorithm:

snap
show
copy

Commands used to specify morphological image transformations in the inspection algorithm are now summarized. Other BLIX commands are further explained in the Watch Gear Inspection Algorithm listing at the end of the paper.

and generates a resultant bit-plane from a pair of input bit-planes, where a given pixel is 1 if and only if the corresponding pixels in both input bit-planes are 1.

or generates a resultant bit-plane from a pair of input bit-planes, where a given pixel is 1 if either corresponding pixel in the input bit-planes is 1.

lgt generates a resultant bit-plane from an ordered pair of input bit-planes, where a given pixel is 1 if and only if it is a 1 in the first input bit-plane and a 0 in the second input bit-plane (Logically Greater Than).

implode generates a resultant bit-plane from a single input bit-plane and an input list of pixel coordinates. The input list gives the x and y locations of the pixels composing the structuring element. The resultant image is 1 at only those pixels marking locations in the image where the structuring element can be centered in such a way that every 1 in the structuring element falls on a 1 in the input bit-plane.

clone generates a resultant bit-plane from a single input pit-plane and an input list of pixels that define a process for creating a structuring element. Clone differs from implode in that all the points of the structuring element need not be defined; only a small subset are required. The clone transformation reconstructs the remainder of the structuring element pixels. Cloning positions a structuring element at every 1 in the input image. The resulting output image is a 1 at those pixels covered by at least one of the cloned structuring elements.

open transforms the specified input bit-plane image by a morphological opening using the specified structuring element. A morphological opening is the image that results when the structuring element is tested at each pixel in the input image for inclusion in the foreground regions (area of 1s) in the input image. The union of all translations of the structuring element that are included in the input image foreground is the opening of the input image. In addition to these commands, BLIX also supports more specialized morphological operators, such as dilation, erosion, and hit-or-miss. BLIX reserves the command *dilation* to mean cloning with a disk or spherical-shaped structuring element.

7. EXAMPLE OF MACHINE VISION INSPECTION FOR MANUFACTURING DEFECTS

This section describes and illustrates an application of morphological image analysis to the inspection of watch gears for missing or broken teeth. The MVI Genesis 2000 is unique in its extensive incorporation of the principles of mathematical morphology as a basis for automated visual inspection and flexible automation control. This example of watch gear inspection is presented to illustrate the significance of the morphological approach to machine vision applications. Further details of the image analysis science of Mathematical Morphology can be found in the text by Serra [20].

The watch gears are shown in Fig. 1 substantially enlarged. Actual size is 0.25 in. in diameter. The backlit gears are imaged using an industrial grade vidicon with macro focusing optics. Resolution is 512 lines × 512 picture

elements per line. The inspection steps described and illustrated here were performed on an MVI Genesis 2000 machine vision development system. The inspection algorithm was programmed in BLIX.
The morphological transformations specified by the BLIX commands composing the Watch Gear Inspection Algorithm implement the following basic image processing steps:

Step 1 Block out the internal holes in the gears.

Step 2 Prune away all the teeth, leaving only a solid round gear body.

Step 3 Construct around the gear body a narrow ring with a diameter just large enough to hit the tips of the gear teeth.

Step 4 Dilate the hit gear teeth tips sufficiently so that normally adjacent gear teeth tips fuse together into a continuous band.

Step 5 Detect any breaks in the band.

Step 6 Denote gear defects in the original image by marking locations where the band is broken.

This inspection algorithm is made particularly difficult because the gears have no center holes and because the purpose of the inspection is to detect broken as well as missing teeth.

Figures 5–16 illustrate the above Inspection Algorithm steps. They were photographed from the color monitor display device that is part of the

Fig. 5 Binary image of the gears to be inspected for missing or broken teeth. The first step of the inspection algorithm is to mask out the holes in the gears, because the holes would otherwise interfere with the remainder of the processing.

Fig. 6 The first stage of the **implode** transformation that locates the centers of the holes in the gears. The **implode** transformation is a basic BLIX command that tests where a structuring element fits into the gear shapes. The structuring element used by the **implode** command in this step is a ring whose diameter is slightly larger than the diameter of the holes.

Fig. 7 A later stage of the **implode** command. The **implode** command is best thought of in terms of two copies of the original gear image sliding across each other, here illustrated in medium and light gray shades. The path followed by the sliding image (medium gray) is circular. As the image slides, any light gray pixels that are not overlapped are removed.

Fig. 8 The result of the **implode** transformation. All the points of the light gray image have been removed except for eight small components that are shown here shifted to the centers of the holes.

Fig. 9 The result of the **clone** operation applied to the light gray bit-plane marking the hole centers. The **clone** transformation in this case specifies that an octagon is to be grown at each hole center pixel.

Fig. 10 The result of the **or** transformation that combines both the light gray and medium gray bitplanes into a single bitplane. At this point step 1 of the inspection.

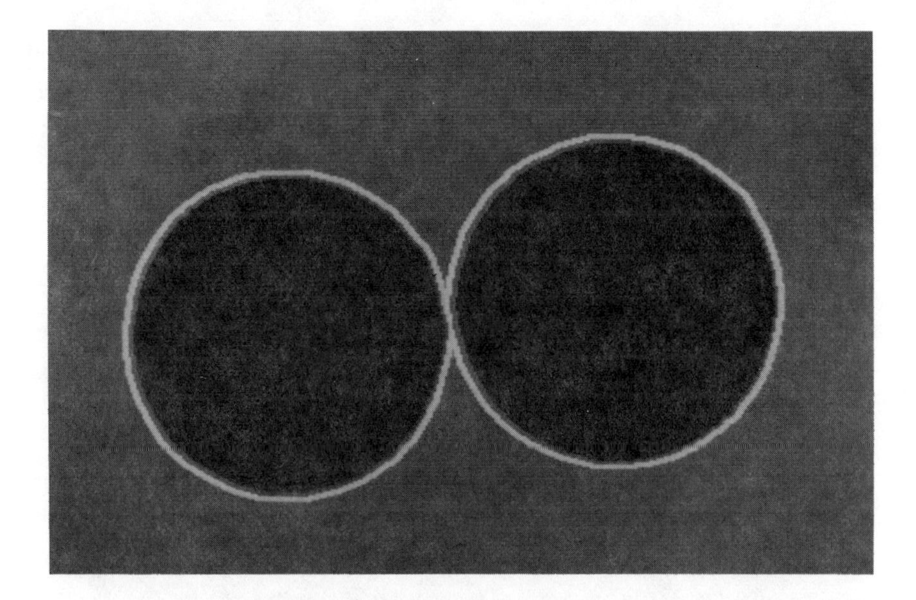

Fig. 11 The result of steps 2 and 3 of the inspection algorithm, the gear pruned of its teeth is shown in black and the sampling ring is shown in white. The toothless gear is the result of the BLIX transformation **open** applied to the bitplane image of Fig. 7 . Here, the structuring element used is a disk whose diameter is slightly less than the diameter of the gear body. The **open** transformation used here has labeled in black only those pixels that would be covered by the structuring element as it slides inside the solid forms of the gears in Fig.10. The preceding step of covering the holes was necessary in order to permit the **open** command to retain the entire gear body. Figure 11 is completed by two **clone** commands that generate first the spacing of the sampling ring away from the gear body and second the sampling ring itself.

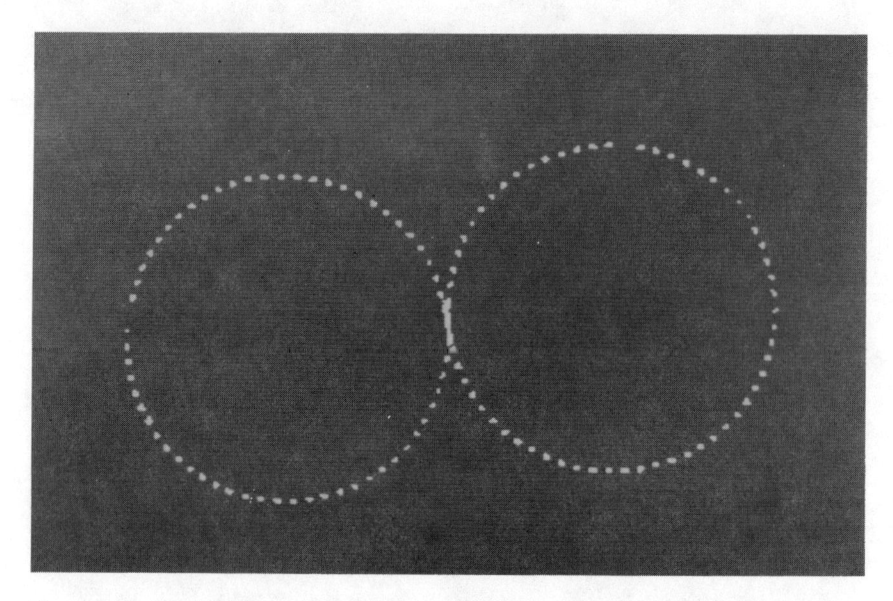

Fig. 12 The gear teeth tips are hit by the sampling ring in white, superimposed on the original image of the gears, and appear here in medium gray.

Fig. 13 The gear teeth tips are imaged by the **and** transformation of the sampling ring bitplane and the original gear image bitplane.

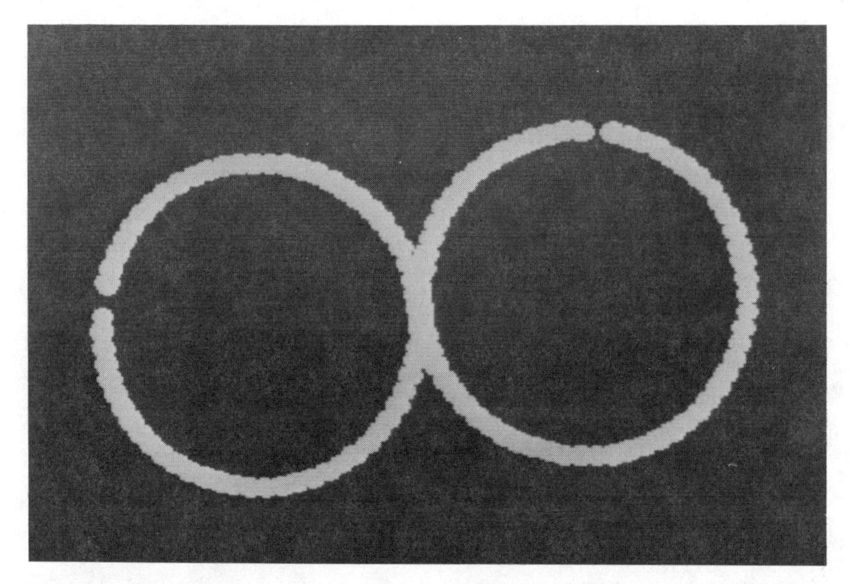

Fig. 14 The results of the **clone** transformation is applied to the gear teeth tips. The structuring element is a disk whose diameter is equal to the nominal gear teeth tip spacing. Where there are no missing teeth, the dilated gear teeth tips fuse together into a continuous band. Where there are gear teeth tips missing, the dilated band is broken.

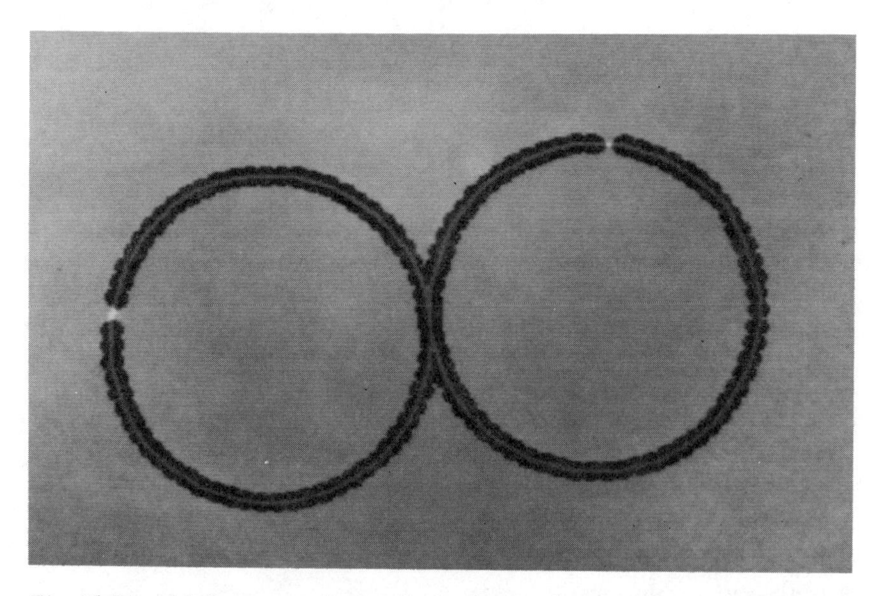

Fig. 15 Two bitplanes are displayed simultaneously: the dilated band and the sampling ring. Where the sampling ring is not overlapped by the band as indicated by the light gray shade, a missing gear tooth tip has been located. Dilating these markers by a disk creates defect cues.

Fig. 16 The original gear binary image is shown with superimposed defect cues.

Genesis 2000 development system. Normally, individual bit-planes are displayed in different colors so that spatial relationships between image features can be observed. In these black-and-white photographs the colors appear as distinct shades of gray.

8. GEAR INSPECTION ALGORITHM LISTING

The following program listing is the complete watch gear inspection algorithm coded in BLIX. An initialization routine for clearing image buffers, loading input and output lookup tables, and filling structuring element data files is followed by a main routine that would be repeated for each new frame of watch gear imagery.
Program lines beginning with a # symbol are treated as comments. BLIX command lines are printed in bold face.

```
#
# Watch gear inspection algorithm—version 1.0.
#
#
# Initialization sequence.
#
```

```
# Clear frame buffers and load standard lookup tables.
#
  init
#
# Display frame buffer 0. on the color monitor using output lookup table.
o0.
#
  show 0. o0
#
# Load the structuring element files with the appropriate data.
# Structuring element generators ring, octagon, and disk generate
# parameters used by the GLU to effect the required morphological
# transformation.
#
  ring 30 16 > hole_ring
  octagon 35 > hole_mask
  disk 75 > gear_body
  disk 3.5 > sampling_ring_spacer
  disk 7.5 > sampling_ring_width
  disk 6.5 > tip_spacing
  disk 10 > defect_cue
#
#
# Inspection algorithm—main routine.
#
# Snap an input image from the camera and place it into buffer 0.
# The digitized video frame is passed through input lookup table i0.
# Input lookup table i0 is assumed to be loaded with the appropriate
# values for thresholding the gray-scale camera output into 1's and 0's.
# The binary gear image is then found in bitplane 0.0
#
  snap i0 0.
#
# Place a copy of the original bit-plane 0.0 image in bit-plane 2.0.
# All image transformations will be done in frame buffer 2.
#
  copy 0.0 2.0
#
# Find the center of the gear holes of bit-plane 2.0 placing the result
# in bit-plane 2.1.
# The structuring element data is in the file hole_ring.
#
```

```
    implode 2.0 2.1 < hole_ring
#
# Construct an octagonal hole mask centered at each nonzero pixel
# of bit-plane 2.1 and place the masks in bit-plane 2.0,
# replacing previous contents.
# Structuring element data is in file hole_mask.
#
    clone 2.1 2.0 < hold_mask
#
# Combine the original gear image with the hole masks, leaving the result
# in bit-plane 2.0.
#
    or 0.0 2.0
#
# Open the hole_masked gears in bit-plane 2.0 by the structuring element
# specified in file gear_body
# to prune away the teeth.
# The resulting gear bodies placed in bit-plane 2.1.
#
    open 2.0 2.1 < gear_body
#
# Clone the gear body in bit-plane 2.1 by a spacing disk whose radius
# is equal to the spacing of the gear teeth tips from the gear body.
# The structuring element data is found in file sampling_ring_spacer
#
    clone 2.1 2.2 < sampling_ring_spacer
#
# Clone the already cloned gear_body image in bit-plane 2.2 by a disk
# whose radius is equal to the width of the sampling ring.
#
    clone 2.2 2.3 < sampling_ring_width
#
# Construct the sampling ring by taking the difference between bit-planes
# 2.2 and 2.3 and place the result in 2.0.
#
    lgt 2.3 2.2 2.0
#
# Find where the sampling ring hits the gear teeth by intersecting
# bit-plane 2.0 with 0.0 and placing the result in bit-plane 2.1.
#
    and 2.0 0.0 2.1
#
```

```
# Clone each gear tooth tip in bit-plane 2.1 by a disk structuring element
# tip_spacing whose diameter is equal to the tip_to_tip gear tooth
# spacing. Place the resulting band in bit-plane 2.2.
#
   clone 2.1 2.2 < tip-spacing
#
# Determine where the sampling ring in bit-plane 2.0 is not covered by the
# band in bit-plane 2.2 by taking set differences and place the result
# in bit-plane 2.3.
#
   lgt 2.0 2.2 2.3
#
# Create gear defect cues by cloning each nonzero pixel in bit-plane 2.3
# by a disk specified in structuring element file defect_cue.
# Place the resulting defect cues into bit-plane 0.1.
#
   clone 2.3 0.1 < defect_cue
#
# The original gear image with overlapping defect cues is now displayed.
#
```

REFERENCES

[1] von Neumann, J. (1966). *Theory of Self-Reproducing Automata*, (A. Burks, ed.). University of Illinois Press, Urbana.

[2] Ulam, S. M. (1957). On some new possibilities in the organization and use of computing machines. *IBM Research Reports*, no. RC68, May.

[3] Unger, S. H. (1958). A computer oriented to spatial problems, *Proc IRE*, **46**, 1744–1750.

[4] Hubel, D. H., and Wiesel, T. N. Receptive fields, binocular interaction and functional architecture in the cat's visual cortex. *J. Physiol.*, 160, 106–154.

[5] Rosenfeld, A. (1983). Parallel image processing in cellular arrays. *IEEE Computer*, **16**, no. 1, January.

[6] Slotnick, D. L., Borck, W. C., and McReynolds, R. C. (1962). The SOLOMON COMPUTER. *Proc. Western Joint Comp. Conf.*, 87–107.

[7] Klein, J. C. and Serra, J. (1973). The texture analyser. *J. of Microscopy*, **95** part 2, 349–356, April.

[8] Preston, K., Jr., (1961). Machine techniques for automatic identification of the binucleate lymphocyte. *Proc. Fourth Int'l Conf. Medical Electronics*, Wash. D.C., July.

[9] Golay, M. J. E. (1969). Hexagonal parallel pattern transformation. *IEEE Transactions on Computing*, **C-18**, 733–740.

[10] Graham, D., and Norgren, P. E. (1980). The diff3 analyzer: A parallel/serial Golay logic processor. *Real Time Medical Image Processing*, (M. Onoe, K. Preston, Jr., and A. Rosenfeld, eds.). Plenum Press, New York.

[11] Sternberg, S. R. (1979). Parallel architectures for image processing. *Proc. IEEE COMPSAC*, Chicago.

[12] Lougheed, R. M., McCubbrey, D. L., and Sternberg, S. R., (1980). Cytocomputers: architectures for parallel image processing, *Proc. Workshop Picture Data Description and Management*, August, pp. 281–286.

[13] Sternberg, S. R. (1983). Biomedical image processing. *IEEE Computer Magazine*, **16**, No. 1, January.

[14] Kung, H. T., and Picard, R. L. (1981). Hardware pipelines for multidimensional convolution and resampling. *Proc. Workshop Computer Architectures PAIDM*, pp. 273–278.

[15] Batcher, K. E. (1980). Design of a massively parallel processor. *IEEE Trans. Comp.*, **28**, 836–840.

[16] Duff, M. J. B., Watson, D. M., Fountain, T. M., and Shaw, G. K. (1973). A cellular logic array for image processing. *Pattern Recognition*, 5, 229–247.

[17] Matheron, G. (1975). Random Sets and Integral Geometry. John Wiley, New York.

[18] Minkowski, H. (1903). Volumen and oberflache. *Math. Ann.*, **57**, 447–495.

[19] Serra, J. (1967). Buts et realisation de l'analyseur de textures. *Revue de l'Industrie Minerale.*

[20] Serra, J. (1982). *Image Analysis and Mathematical Morphology*. Academic Press, London.

[21] BLIX Reference Manual, version 1.0. July, 1983. Machine Vision International, Corp.

[22] Sternberg, S. R. (submitted for publication). Grayscale morphology. In *Computer Graphics and Image Processing.*

Chapter Seven

*Hierarchical Line Linking for Corner Detection**

Ralph Hartley and Azriel Rosenfeld

1. INTRODUCTION

When one is matching images taken at different times or from different view directions, it is often useful to segment the images into regions representing objects in the real world and to match the prominent features of these regions. Corners on the region boundaries are one good set of features to match because they are relatively invariant under changes in perspective.

Many authors, e.g., Freeman and Davis [1] and Rutkowski [2], have found corners by doing local operations on a chain-coded description of the boundary. A major problem with such approaches is that it is difficult to determine in advance how large a neighborhood is required for the operations, i.e., to determine the scale of the corners that are to be detected. This problem is exacerbated by the fact that a boundary can simultaneously contain corners at different scales, as illustrated by Fig. 1. It is desirable to find both the fine-scale corners and the coarse ones. Another problem is that if the neighborhood is too large, then the computational cost can also be large.

Fischler [3] finds a point that is at a maximum distance from a chord of the curve and marks it as a corner if the curve makes exactly one significant excursion away from the chord. Chords of several different lengths are used to find corners at different resolutions. This is done at every possible position on the curve. The minimum displacement from the chord at which a point is considered to be a corner candidate depends only on the length of the chord;

* The support of the National Science Foundation under Grant MCS-82-18408 is gratefully acknowledged, as is the help of Janet Salzman in preparing this paper.

INTEGRATED TECHNOLOGY FOR PARALLEL IMAGE PROCESSING
ISBN 0-12-444820-8

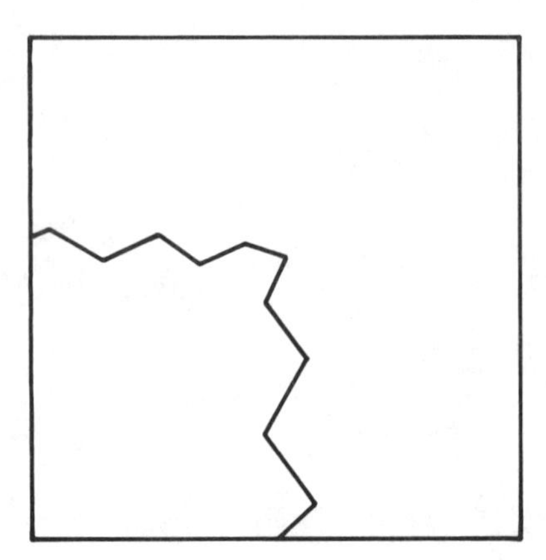

Fig. 1 Curve having corners at two different scales

therefore, corners that have small angles (close to 180 degrees) but are between two long straight stretches of curve are not found, even though they are perceptually clear.

Langridge [4] iteratively fits a cubic curve to a sequence of points and then detects corners at maxima of a local measure of curvature. The x and y components of the curve are handled separately, which causes the results to be dependent on orientation. It is also unclear what happens when the results from analysis of the two coordinates give different positions for a corner. Because the curve is smoothed over a neighborhood containing only five points, the number of iterations required increases rapidly as the length of the curve increases.

Langridge [5] and O'Callaghan [6] compute properties of convex contours on neighborhoods the size of which depends on the shape of the curve, so that larger neighborhoods are used on straight sections of the boundary. Several criteria are used to find points that dominate a neighborhood. These points do not always correspond to what we think of as corners.

The solution presented here to these problems uses a hierarchy of images at different scales. Such a structure, called a pyramid, has several advantages over a regular array. In a pyramid, global (or large-scale) properties of figures can be found by purely local operations because at some level, any region of the image is represented by a small set of cells. If each cell of a pyramid is connected to its neighbors on the same level and on the levels above and below, then any two cells are connected by a path that is at most $2 \log n$ long. This contrasts with a regular array with neighborhood connections, in which minimum paths can be up to n long. As a result, many operations that require $O(n)$ time in a regular array can be done in $O(\log n)$ time in a pyramid. This is especially true if parallel hardware is available. Such hardware is currently being developed [7].

Pyramids have been used for recognizing features in gray-level images [8]. The basic idea is to apply a local operation to the pyramid in which each cell builds a summary of the information in the cells that it represents in the next finer image. A mechanism for producing a strip-tree-like representation of a curve [9] was described in Hong *et al.* [10]. The approach used here is similar. Curves are approximated by line segments. At each level of the pyramid, line segments from the level below are combined and examined to find the corners at the scale represented by that level.

2. METHOD

The algorithm uses a pyramid in which each level has half the linear dimensions of the level below it. Each node represents a 4 × 4 region on the level below it, but for adjacent nodes these regions overlap by 50%, as illustrated in Fig. 2. This means that each node is represented by four nodes on the level above. Without this overlap, features that cross the boundaries of

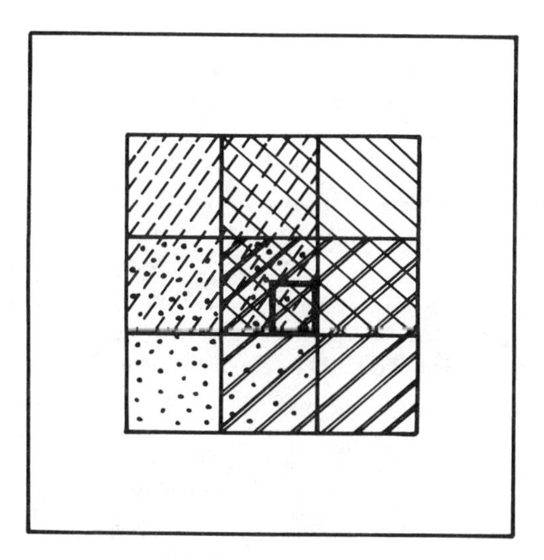

Fig. 2 Areas on level $n + 1$ covered by overlapping blocks of nodes on level n. (The small square with the heavy outline is a level-n node.) The blocks are denoted by dots (lower left), dashed lines (upper left), solid lines (upper right), and double lines (lower right).

high-level nodes could fall between the cracks. Each node contains a description of any curves that pass through the region of the image that it represents. A curve is represented by a line segment and an estimate of the amount of deviation from the line segment (the mean squared error). Note that although the higher levels contain a coarser description of the curves, in the sense that they are represented by fewer line segments, the line segments themselves are stored at the resolution of the original image. For each line segment, the sums (over the curve represented by the segment) Σx, Σy, Σx^2, Σxy, and Σy^2, as well as k, the amount of support the segment receives from the original image, are stored. These values are computed hierarchically, and the parameters of the line segments are computed from them.

The input to the program is a $2^n \times 2^n$ image that has been processed by a local edge or line detector, so that each edge point is marked with a direction. This input is copied into the bottom level of the pyramid by assuming that each edge is a directed line that passes through the center of the pixel and passes all the way across the square represented by the pixel. The lines can be weighted by the strength of the edge detection.

The program makes a single pass through the pyramid from bottom to top. Each node is processed independently of all other nodes on the same level; th the processing could be done in parallel on a cellular pyramid machine. Each node examines the curves in its corresponding 4×4 region on the level below. Curves that are continuations of each other (as determined by criteria to be described below) are grouped together and are combined into single curves. Curves that are too short to be significant at the current level are then discarded to keep the number of curves to be dealt with at each node reasonably small. Curves are evaluated by other criteria to find those that form corners. Corners are found independently on each level.

When curves are grouped together, the values of Σx, Σy, Σx^2, Σxy, Σy^2, and k are combined. These values contain enough information to compute a least-squares fit to obtain a new approximation by a line segment. The fit minimizes the sum (over all the points on the curves being combined) of the square of the perpendicular distance to the new line. The method used to do the fit is described in Pavlidis [11]. The new line has slope defined by the equation

$$\tan 2\Theta = \frac{2(\overline{xy} - \bar{x}\,\bar{y})}{(\overline{y^2} - \bar{y}^2) - (\overline{x^2} - \bar{x}^2)}$$

and passes through the point $(\bar{x}, \bar{y})$. The error is given by the formula

$$D^2 = (\overline{y^2} - \bar{y}^2)\cos^2\Theta - 2(\overline{xy} - \bar{x}\,\bar{y})\sin\Theta\cos\Theta + (\overline{x^2} - \bar{x}^2)\sin^2\Theta$$

where

$$\bar{x} = \frac{\Sigma x}{k} \qquad \bar{y} = \frac{\Sigma y}{k}$$

$$\overline{x^2} = \frac{\Sigma x^2}{k} \qquad \overline{xy} = \frac{\Sigma xy}{k} \qquad \overline{y^2} = \frac{\Sigma y^2}{k}$$

Because the input data consist of line segments instead of discrete sets of points, Σx, Σy, Σx^2, Σxy, and Σy^2 (for nodes on the bottom level of the pyramid) are actually integrals over the length of the line segments. If (x_0, y_0) and (x_1, y_1) are the coordinates of the end points of a line segment, then the sums for the segment are determined by

$$\bar{x} = \frac{x_0 + x_1}{2}$$

$$\bar{y} = \frac{y_0 + y_1}{2}$$

$$\overline{x^2} - \bar{x}^2 = \frac{(x_1 - x_0)^2}{12}$$

$$\overline{xy} - \bar{x}\bar{y} = \frac{(x_1 - x_0)(y_1 - y_0)}{12}$$

$$\overline{y^2} - \bar{y}^2 = \frac{(y_1 - y_0)^2}{12}$$

because each node on the bottom level represents a single pixel $k = 1$.

For nodes on all levels above zero, Σx, Σy, Σx^2, Σxy, Σy^2, and k are calculated hierarchically. When two curves are combined, the sums for the new curve are weighted averages of those of its parts. The contribution of a curve is weighted according to its position relative to the node, as described in the discussion. That this gives the correct values for the entire curve, regardless of how the curve is broken up on lower levels, is shown in the next section.

For each curve, the two extremal points in the original image are found. This is also done hierarchically by starting, on level zero, with the end points of the initial curves (which are line segments). When curves are combined, the new extremal points are the two that are extremal in the direction of the new line segment. These points are used to determine the end points of the new curve and also make it possible to localize corners exactly.

3. DISCUSSION

A desirable property of hierarchical systems is that the result not be drastically distorted by the structure used in the computation. In the case of our algorithm, it is particularly important to show that the way in which curves are combined and summarized does not cause artifacts in the results. One would like to show that a single curve, which is recognized as such by the evaluation criterion, will be handled correctly. This means that the node representing the entire curve should contain the same description of the curve that would have been given by a nonhierarchical method and that the description should be invariant with respect to the position of the grid of the pyramid relative to the figure.

The representation of a curve in our algorithm has two main parts: the sums that are used to fit the line and the two end points. All other information about a curve is derived from these. It will be shown here that the sums satisfy the constraints described above and that the end points violate them only to a small degree.

The first constraint, that the representation of the curve at the apex of the pyramid be the same as a similar representation computed nonhierarchically, is satisfied by the sums because every node makes the same total contribution to the level above. Let Σx, Σy, Σx^2, Σxy, Σy^2, and k for node n on level L be denoted by the six-component vector $S_L(n)$. Also, let $N(L)$ denote the set of nodes on level L that represent the curve. Because the pyramid tapers exponentially, the set $N(L)$ becomes smaller at each level until it has just four members (the curve is always redundantly represented by at least four nodes because of overlap in the pyramid). The globally computed values of the sums for the entire curve are, by definition, the values of the sums summed over the set of nodes that represent the curve at level zero (the input image). That is,

$$S_{\text{global}} = \sum_{n \in N(0)} D_0(n)$$

The sums $S_{L+1}(n)$ for each node n in $N(L + 1)$ are computed by taking a weighted sum over a neighborhood on level L. Thus

$$S_{L+1}(n) = \sum_{m \in N(L)} w(n, m) S_L(m)$$

where $w(n, m)$ depends on the relative location of n and m, and is zero if m is not within the 4×4 neighborhood on level L represented by n. The actual values of the weights are determined by the second constraint. The first constraint requires that

$$S_{\text{global}} = \sum_{n \in N(L)} S_L(n)$$

for all L. If, when nodes from level L are combined to form the nodes of level $L + 1$, each node m on level L has the same total weight, that is if

$$\sum_{n \in N(L+1)} w(n, m) = c$$

and if the weights are normalized so that $c = 1$, then the sum of all the nodes on level $L + 1$ is the same as the sum of the nodes on level L, because

$$\begin{aligned}
\sum_{n \in N(L+1)} S_{L+1}(n) &= \sum_{n \in N(L+1)} \sum_{m \in N(L)} w(n,m) S_L(m) \\
&= \sum_{m \in N(L)} \sum_{n \in N(L+1)} w(n,m) S_L(m) \\
&= \sum_{m \in N(L)} S_L(m) \sum_{n \in N(L+1)} w(n,m) \\
&= \sum_{m \in N(L)} c S_L(m) \\
&= \sum_{n \in N(L)} S_L(m)
\end{aligned}$$

Therefore, by induction, the total sums for each level are the same as the globally computed sums. That is,

$$S_{\text{global}} = \sum_{n \in N(L)} S_L(n)$$

for all L. When the sums are concentrated into one node by the tapering of the pyramid, that node then contains the global sums. Figure 2 illustrates that all nodes do, in fact, contribute equally to the level above. Each node contributes to four cells: to the center of one cell, to the edges of two cells, and to the corner of one cell. As long as the weights are symmetric about the center of the region represented by a node, then the weight of a node on the level below depends only on which of the three cases described holds. Therefore, each node contributes to the same number of nodes on the level above, with the same set of weights, and the requirement is satisfied. This fact was noted by Burt [12]. Note that this is a special property of the 4×4 neighborhood with a factor-of-two reduction at each level; for other arrangements the cells must be weighted properly, according to their positions relative to the cell, to satisfy the equal contribution requirement. This is also not the case for a node on the edge of the array, but it is a simple matter to give such nodes extra weight to compensate for this.

The second constraint requires that the information on each level not depend on the position of the figure relative to the grid. This requires that only low frequency components of the shape of the curve be passed to the higher levels of the pyramid, so that the shape is sampled sufficiently. The maximum amount of information about the shape can be passed up, while maintaining consistency with this constraint, by convolving the input (S_0) with an approximation of a Gaussian, the standard deviation of which depends on the level. This is equivalent to passing the data through a low-pass filter.

The sums of each node n on level L are weighted averages of the sums on level 0:

$$S_L(n) = \sum_{m \in N(0)} W_L(n, m) S_0(m)$$

where $W_L(n,m)$ is a function of the position of m relative to n. Because

$$\begin{aligned} S_L(n) &= \sum_{p \in N(L-1)} w(n,p) S_{L-1}(p) \\ &= \sum_{p \in N(L-1)} w(n,p) \sum_{m \in N(0)} W_{L-1}(p,m) S_0(m) \\ &= \sum_{m \in N(0)} \sum_{p \in N(L-1)} w(n,p) W_{L-1}(p,m) S_0(m) \end{aligned}$$

W_L is determined by the formula

$$W_L(n,m) = \sum_{p \in N(L-1)} w(n,p) W_{L-1}(p,m) \qquad \text{for } L > 1$$

$$W_1(n,m) = w(n,m)$$

W_L can be made to approximate a Gaussian if the local weights w are chosen as shown in Fig. 3. That this is so and that it results in adequate sampling are shown by Burt [12]. Gaussian weights are desirable because a Gaussian has a

	.13	.37	.37	.13
.13	.0169	.0481	.0481	.0169
.37	.0481	0.1369	0.1369	.0481
.37	.0481	0.1369	0.1369	.0481
.13	.0169	.0481	.0481	.0169

Fig. 3 Weights used to compute a Gaussian pyramid

good space-bandwidth product. If too much weight is put on the center sons, the sharp cutoff of W_L will produce aliasing, causing the difference between lines represented by adjacent nodes to depend on the position of the figure relative to the grid. This could result both in detection of false corners in smooth curves and in failure to detect corners that are present. If too little weight is put on the central sons and too much on the outer ones, the shape will be unnecessarily blurred on the upper levels, and some corners could be lost.

The end points of the curve do not, in general, satisfy the first constraint exactly; they are not exactly the same as the points that would be given by a global search. They are also more difficult to analyze formally. Figure 4 illustrates how the constraint can be violated. Here A and B are two curves on level L with end points 1, 2, and 3. If these lines are combined, on level $L + 1$, to form curve C, then C will have 1 and 3 as end points. The globally best end point, 4, is not found because it is not the end point of one of the parts. This will not happen often except for very tortuous curves. Because there must be a sharp turn for this problem to appear, curves that have the potential of causing the problem will often be treated as forming a corner. The second constraint is satisfied by the end points, ignoring the effect described above, because the end points are computed from the entire portion of the curve that contributes to a node. If the end of the curve is in

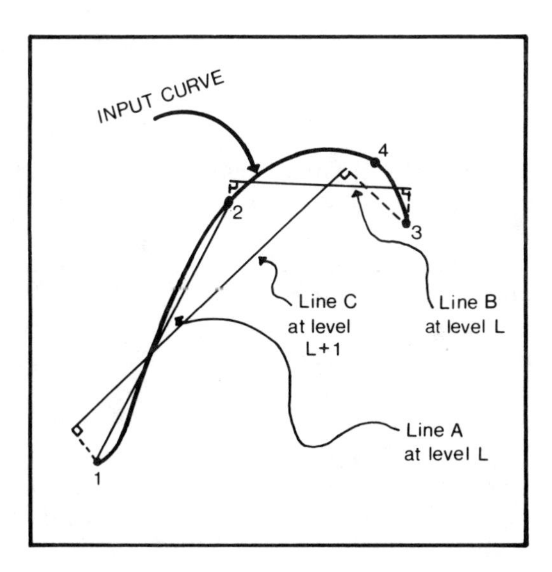

Fig. 4 The end points of a combined curve may not be the same as the end points of its parts

this area, then the end point is the end of the curve; otherwise, the end point is the point at which the curve leaves the area. The computation can suffer, however, from the fact that the end points can, and often will, come from parts of the curve that have little or no weight in computing the approximation. None of these problems are serious if the curve does not change too much in orientation over its length.

4. CRITERIA

The procedure used to group curves depends on exactly what sort of corner-like feature is to be found. Two different sets of criteria are described in this section. The first set of criteria finds corners, on a given level, if the corner is large enough in scale to be significant at that level and if it can also be localized on the lower levels. This definition only allows corners that are sharp; rounded corners will not be found. The other set of criteria defines a corner as an angle flanked on both sides by long, straight sections of curve. The longer and straighter the curve on both sides, the smaller the angle that can be considered a corner.

The first grouping procedure is conservative in that it is designed never to group curves that could define a corner at any higher level. This is possible because of the requirement that a corner have support from the lower levels. Because of this and because of the fact that the regions represented by adjacent nodes overlap, it is often possible to combine curves without evaluating the good continuation criterion. Figure 5 illustrates this possibility. Here a, b, and c represent line segments two levels below the one we are considering. At the next level, a and b were merged to form segment d whereas b and c were merged to form segment e. Because d and e represent the same data from the original image over part of their length, if the curves are separated by a corner, then at least one of them must cross a corner; i.e., there must be a corner either between a and b or between b and c. But because it is assumed that corners were not missed at the level below, it is always safe to combine curves that overlap in this way. This means that a corner, to be detected, cannot occur in a section of curve that is locally smooth; it must be possible to localize the corner to a point.

The criterion used to decide whether or not to combine two curves is illustrated in Fig. 6. $Line_1$ and $line_2$ represent two line segments from the level below that are being considered for merging. $Error_1$ and $error_2$ are the errors in the approximations of the underlying curves by the line segments. Some previously used collinearity measures [13] are not suitable, because they break down if the line segments overlap, and because they do not take into account the goodness of fit (i.e., the degree of straightness) of the

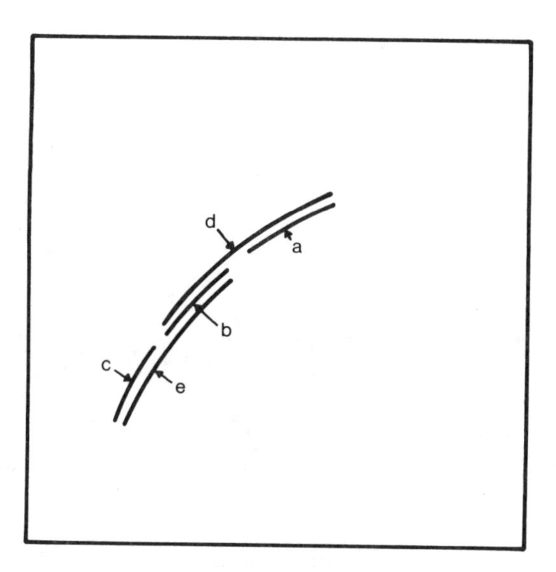

Fig. 5 Overlapping curves

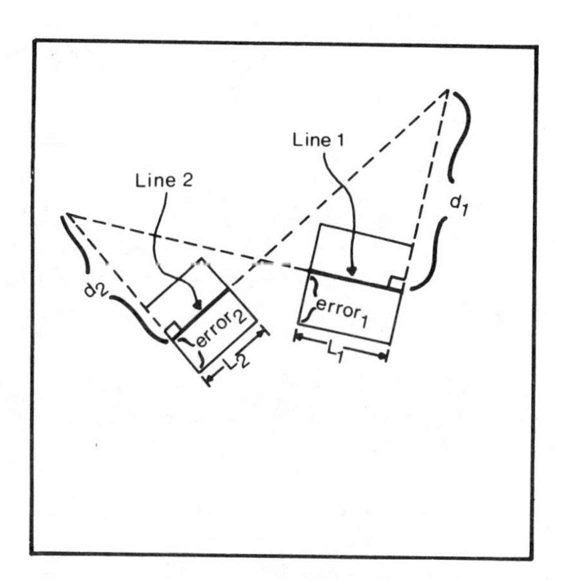

Fig. 6 Error measures

segments. Consideration of the goodness of fit is desirable because it is reasonable to require that two perfectly straight lines line up exactly, but straight-line approximations of curves cannot be expected to be exactly collinear. The maximum of $d_1/error_1$ and $d_2/error_2$ is used as a measure of how well the two curves can be approximated by a line. This measure is computed for each pair of curves visible from the node. The pair for which the measure is minimum and less than a threshold (3.0 in all the examples shown—the exact value of this parameter did not seem to affect the results significantly) is merged until the measure is above the threshold for all pairs. If the error associated with a curve is smaller than a tolerance, which depends on the length of the curve (and hence on the level in the pyramid), then the tolerance is used instead of the error (in our experiments the tolerance was one-fourth of the length of the curve). This allows curves that could not link at a fine level to be combined at a coarse level. Curve segments whose approximating line segments form acute angles are not allowed to link to each other.

The criterion used for finding corners is similar except that the two curves are required to meet. If, after all possible pairs have been merged, there is a pair for which the measure defined above is very large (twice the threshold), and one of the extremal points of one curve is near (within a pixel or two) an extremal point of the other, then the point midway between these two extremal points is marked as a corner.

The second set of criteria combines any line segments that are not separated by an acute angle or a right angle. Lines separated by an acute angle are not combined because when an acute angle is found, a corner is assumed to be present. Pairs of segments that are combined into the same line are examined to find corners. The degree of cornerity is measured by the ratio Θ/F, where Θ is the angle between the segments and

$$F = \max\left\{\frac{error_1}{L_1}, \frac{error_2}{L_2}\right\}$$

(see Fig. 6). F is a measure of how straight the two segments are. Because both F and Θ are dimensionless quantities, the criterion does not depend on scale.

When corners are detected by finding pairs of segments for which this measure is large, a given corner in the original image can be detected more than once. This situation is shown in Fig. 7, in which all the segments are straight, and segments A and B are at acute enough angles to C and D for a corner to be detected. As a result, the corner at 1 is found 4 times: at the points 1,2,3, and 4. Note that segments B and C are closer to the corner, and therefore, may give more accurate information on the corner's location. It does no good to ignore the corners on this level and assume that they will be detected on the next level, after A has been merged with B and C with D,

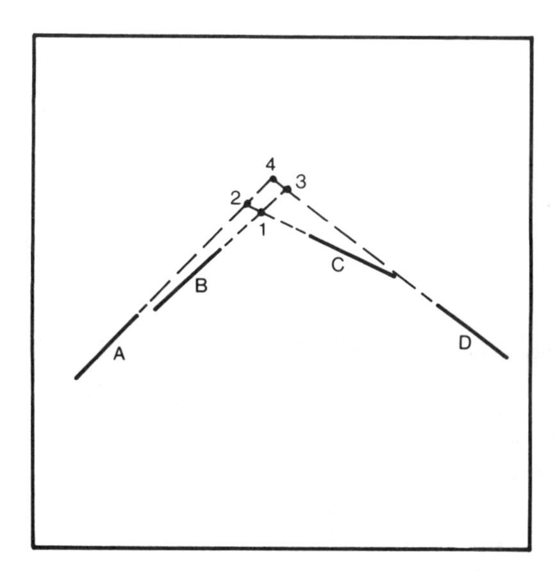

Fig. 7 The same corner may be detected many times

because B and C may also be merged, blurring the position of the corner even more. Also, A and D may be merged with other segments, so that they are no longer sufficiently straight. Another possible problem is illustrated by Fig. 8. Here there are actually three corners, but a fourth corner, located at point 4,

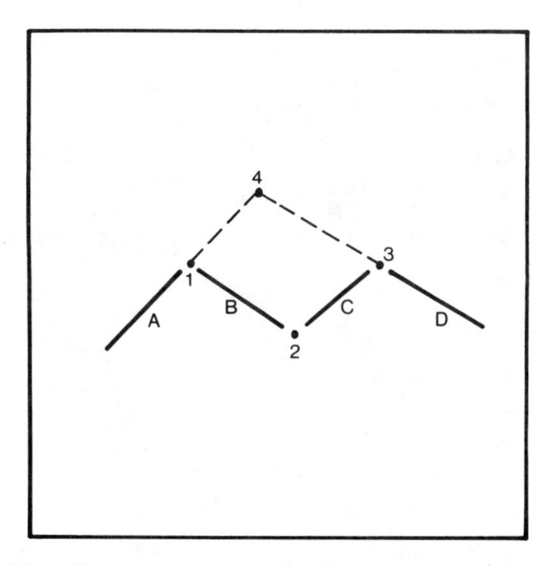

Fig. 8 Nonadjacent segments could give rise to a spurious corner

is found. To prevent this kind of problem the segments are sorted according to their positions along the line. Corners are detected only when a pair of segments forms a corner, and no other pair of segments between them forms a corner.

5. EXAMPLES

Figure 9 shows the corners found, by the first set of criteria, at levels 2 and 5 of the pyramid for an input similar to that of Fig. 1. At one level the program has found the small-scale corners, and at a higher level it found the one big corner.

Figures 10 and 11 are reproduced from Rutkowski [2]. They show an irregular contour and the points where human subjects perceived corners. Figure 12 shows corners detected by the first set of criteria on levels 1 through 4. Almost all of the points selected by human subjects as places where the curve "turns" were marked as corners on at least one level. The most obvious problem, the corner on the extreme left edge of the figure, which is not found on higher levels even though it is a large-scale corner, is due to the fact that the corner has an extremely obtuse angle and, therefore, looks almost like a straight line at the low levels of the pyramid. The first set of criteria are not

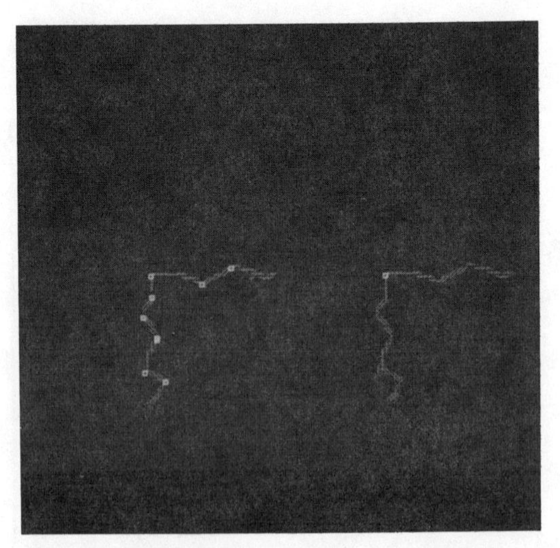

Fig. 9 Corners found on Fig. 1 at levels 2 and 5

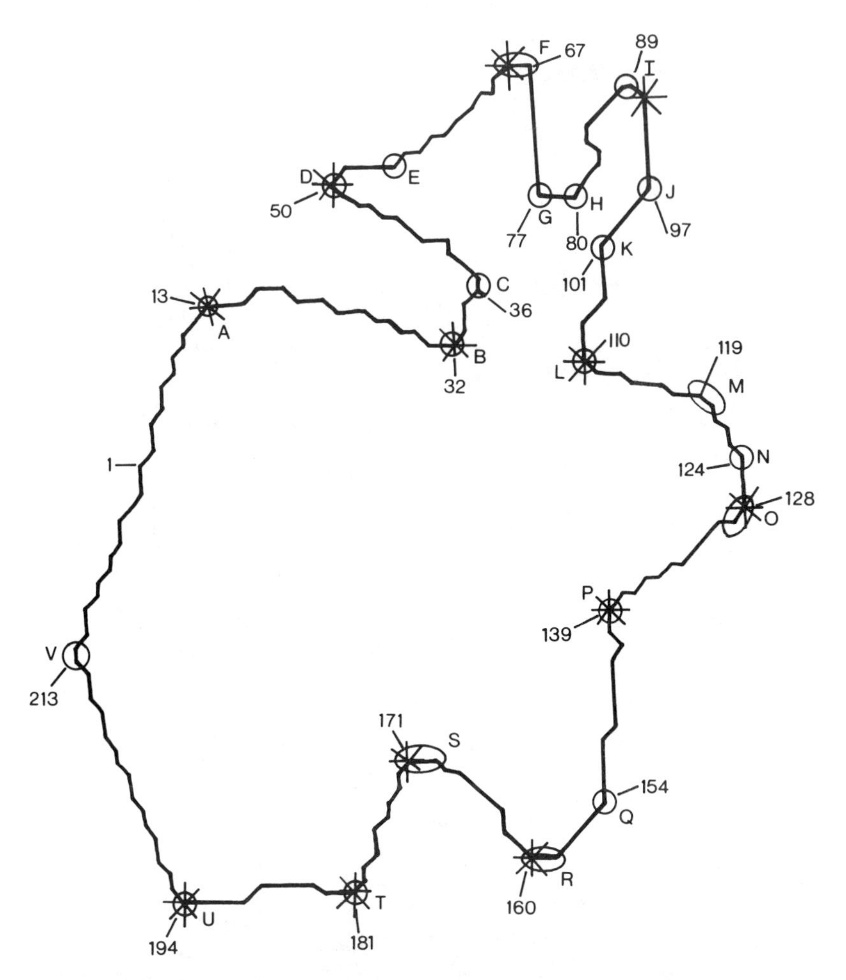

Fig. 10 Curve used in Rutkowski [2] and Fischler [3]. The letters A,. . . V correspond to the letters along the x-axis in the plots shown in Fig. 11. The numbers are link numbers in the chain code. The stars are the points at which the algorithm of Fischler found corners

able to correct this mistake at higher levels. Corners with very obtuse angles are not found even though they are perceptually significant.

Figure 13 shows the same results using the second set of criteria. Note that the problem with large obtuse corners does not occur. Because this method detects blunt corners, for which the exact location is undefined, corners are located with less precision in the higher levels of the pyramid. The apparent false corner just below the obtuse corner at the left edge of the picture is

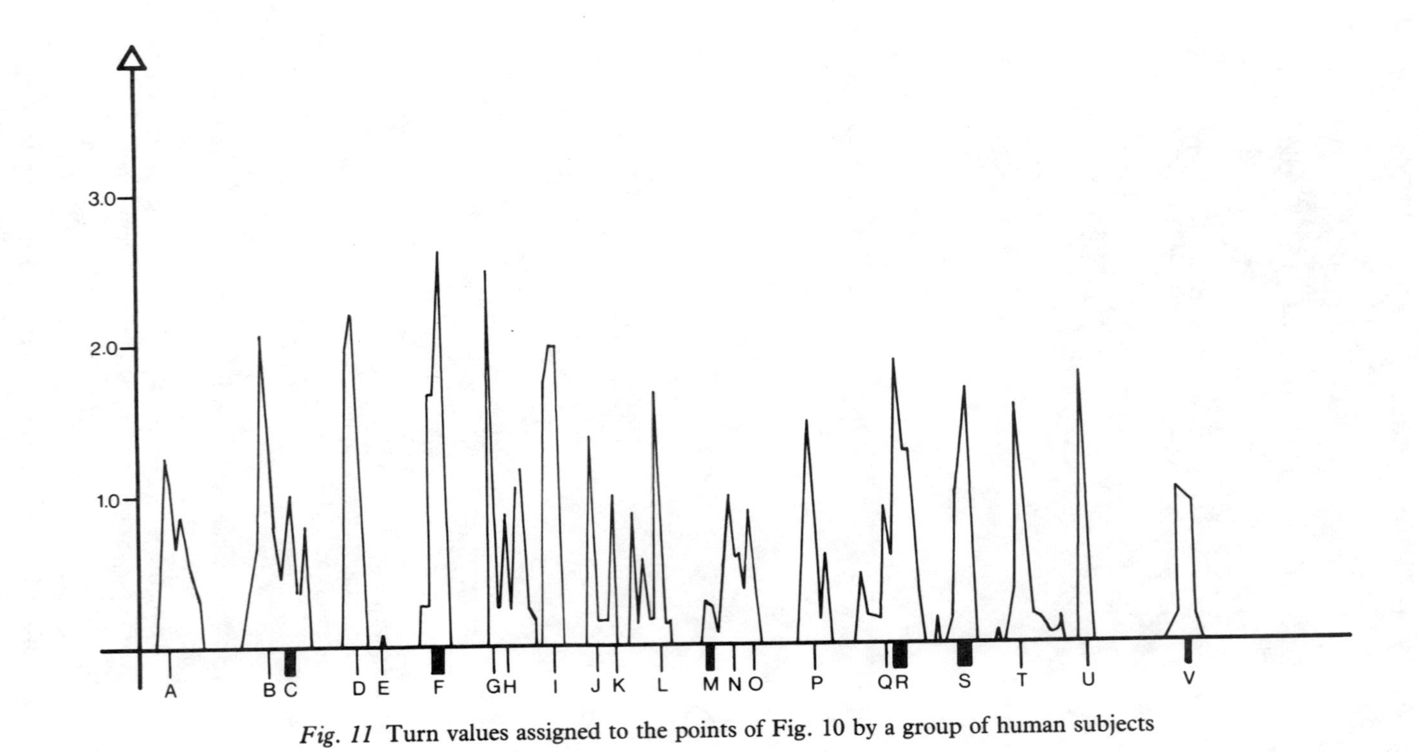

Fig. 11 Turn values assigned to the points of Fig. 10 by a group of human subjects

Fig. 12 Corners found on Fig. 10 at levels 1–4 using first algorithm

Fig. 13 Corners found on Fig. 10 at levels 1–4 using second algorithm

actually a detection of the same corner for which the position is incorrect. The results of both sets of criteria were comparable to those of Fischler and Bolles [3] (whose results for the same curve are marked "*" in Fig. 10), which not only did not detect the corner near the left edge but also were overwhelmed by the complicated region near the top of the figure.

6. CONCLUSIONS

Features are often present in an image at many different scales. Such features can, in principle, be detected directly from the original image. These methods must examine the image globally if they are to find the large-scale features. This often requires a large amount of computation. On the other hand, this global processing is unnecessary when finding small-scale features. This suggests that it might be desirable to use only local information to detect small-scale features and to use global information only to detect large features. If large features can be detected from a coarse summary of the original data, then it is possible to build a hierarchy of representations and detect features of all sizes using local operations within this hierarchy.

This approach has been applied to the problem of finding corners in contours. The program builds a hierarchical representation of a contour by fitting longer and longer line segments to the data and then finds corners by examining small neighborhoods in this representation. The results compare favorably with other methods of detecting corners. It is planned to apply similar methods to finding other features such as pairs of antiparallel lines or "generalized ribbons."

REFERENCES

[1] Freeman, H., and Davis, L. (1977). A corner-finding algorithm for chain coded curves. *IEEE Trans. on Computers*, **26**, 297–303.

[2] Rutkowski, W. (1977). A Comparison of Corner-Detection Techniques for Chain-Coded Curves, University of Maryland, TR-623, College Park, Maryland.

[3] Fischler, M., and Bolles, R. (1983). Perceptual organization and curve partitioning. *Proc. of the IEEE Computer Society Conference on Computer Vision and Pattern Recognition*, 38–46.

[4] Langridge, D. (1982). Curve encoding and the detection of discontinuities. *Computer Graphics and Image Processing*, **20**, 58–71.

[5] Langridge, D. (1972). On the computation of shape. In *Frontiers of Pattern Recognition* (*S. Watanabe, ed.*). Academic Press, New York, pp. 347–365.

[6] O'Callaghan, J. (1974). Recovery of perceptual shape organizations from simple closed boundaries. *Computer Graphics and Image Processing*, **3**, 300–312.

[7] Tanimoto S. (1982). Programming techniques for hierarchical parallel image processors. In *Multicomputers and Image Processing—Algorithms and Programs* (*K. Preston, Jr. and L. Uhr, eds.*). Academic Press, New York, pp. 421–420.

[8] Hong, T. (1982). *Pyramid Methods in Image Analysis*, Ph.D. Thesis, University of Maryland. College Park, Maryland.

[9] Ballard, D. (1981). Strip trees: a hierarchical representation for curves. *Communications of the ACM*, **24**, 310–321.

[10] Hong, T., Shneier, M., Hartley, R., and Rosenfeld, A. (1983). Using pyramids to detect good continuation. *IEEE Transactions on Systems, Man and Cybernetics*, **13**, 631–635.

[11] Pavlidis, T. (1977). *Structural Pattern Recognition*, Springer-Verlag, New York.

[12] Burt, P. J. (1981). Fast filter transforms for image processing. *Computer Graphics and Image Processing*, **16**, 20–51.

[13] Broder, A., and Rosenfeld, A. (1981). A node on collinearity merit. *Pattern Recognition*, **13**, 237–239.

Chapter Eight

A Pyramid Project Using Integrated Technology

V. Cantoni, M. Ferretti, S. Levialdi, and F. Maloberti

1. INTRODUCTION

Various investigators have explored two main approaches to fast processing of images, particularly for low-level vision applications: the array of processors and the pipeline of processors. (For a discussion of the relative merits of these approaches see Cantoni and Levialdi [1]). Recently, a proposal for merging the advantages of both has been suggested by a few authors (for a basic reference book see Tanimoto and Klinger [2]), which is generally known as the *hierarchical computation structure*.

Alternative names for these architectures are *perception cones*, introduced by L. Uhr [3] who is particularly interested in developing models of perception, *pyramids* introduced by C. R. Dyer [4] and others and originating from their interest in recursive data structures (quadtree, octree, and pyramid) [5], [6], [7].

A pyramid structure can be considered also as the three-dimensional extension of the two-dimensional binary tree (Fig. 1). Three features of this structure suggest its usefulness in image processing. The first is related to the well-known order logarithmic dependence for the interprocessor communication supported by this topology; the second exploits interplane communication for the implementation of a planning strategy using images at different resolution levels [7], [8]; the third depends on the possibility of different images flowing towards the apex in a pipeline mode.

To exploit fully the above advantages, we have designed a multiprocessor pyramid architecture, Pyramidal Architecture for Parallel Image Analysis

INTEGRATED TECHNOLOGY FOR PARALLEL IMAGE PROCESSING
ISBN 0-12-444820-8

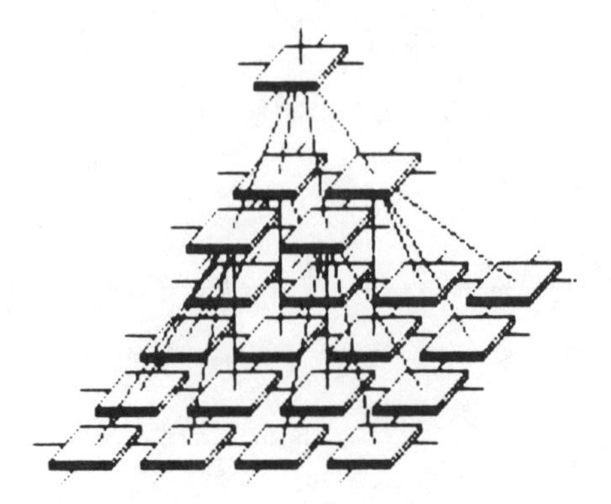

Fig. 1 Processor interconnection scheme of the pyramidal structure

(PAPIA), made of tapered layers of processors, each layer being a truly SIMD machine. Moreover, different layers can execute different instructions, thus becoming a Multi-SIMD processing system, or alternatively, a given subset of layers can operate in the SIMD mode.

The single processor is horizontally 4-connected to its brothers on the same layer and vertically to a top father and to four bottom sons on its higher and lower layers, respectively. Border elements (on the four corners or on the sides) assume that the values of nonexisting neighbors are 0 by default or 1 when specified.

The image loading is overlapped with the processing using column parallelism; i.e., all values that correspond to pixels belonging to the same column are entered simultaneously. All intercommunication buses are single bit, and the processor has a one bit ALU implementing serial arithmetic. An overall view of the features embedded in the processor is given in Section 2.

Since this will be a prototype machine, and this project is a testbed for a number of ideas on a best architecture for image processing, a decision was made to avoid the use of VLSI circuitry at its maximum degree. The layout of the circuit is currently being manufactured with a 4-μm Si-gate NMOS technology.

Section 3 includes details of the chip layout; some choices were made on the basis of the manufacturer's requirements, particularly severe in connection with chip area (5×5 mm^2) and the chip package (a standard unit using 48 pins dual-in-line).

Section 4 contains a perspective view of future developments and some conclusions regarding possible fields of application.

2. THE PROCESSING ELEMENT

The majority of image processing systems based on a multiprocessor architecture operating in the SIMD mode use one-bit processors, which are particularly suitable for the execution of boolean operations on binary images (as is well known, grey-level images are represented by $\log_2 n$ binary planes, where n is the number of grey levels) [9], [10]. Following this trend we have also used a one-bit processor [11], [12], which can be seen schematically in Fig. 2; the function of each of the labeled boxes will be summarized. The local memory of the processor is external to this diagram and contains 256 bits of static RAM storage.

Two main registers, shown as A and B, contain the present status of the corresponding cell and the result of the near-neighbor function of the inputs coming from the adjacent neighbors, respectively. Conversely, the B register may be by passed using the S1–S8 path.

The ALU is composed of a boolean processor, a full adder with an extra register C for the carry bit, and a comparator between the contents of the two shift registers Sr1 and Sr2. All registers, except the shift registers, are single-bit master/slave registers. Sr1 and Sr2 have a variable length with a maximum of 32 bits; their length can be programmed depending on the number of bits per pixel of the image. Arithmetical operations are performed in Sr1 and Sr2; the results are placed in Sr2 (accumulator). Convolutions, for instance, can be achieved without requiring the storage of partial results in the local memory, because Sr2 acts as a buffer memory, and during the multiplication phase Sr1 cyclicly presents the value to be multiplied.*

Both horizontal and vertical connections are provided, but they cannot operate simultanously: five inputs are supplied so that four of them work on the processor layer (along the four cardinal directions from its brothers), or four plus one on the lower and upper layer, respectively (toward the four sons and the father). The near-neighbor function is implemented by the gating technique; and by using S1 and S6, four boolean functions can be chosen to be applied on the selected subset of neighbors.

When considering typical tasks of image processing performed on SIMD machines, one can see [13] that most of the computation time is used in data exchanges (communication) between processors. Moreover some algorithms require local operations on neighborhoods larger than those circumscribed by 3×3 windows. Some ad hoc hardware solutions have been introduced in a few SIMD machines [9], [12] to allow faster communication between

* To speed up the multiplication execution, Sr1 is one cell longer than Sr2, allowing efficient implementation of the serial arithmetic.

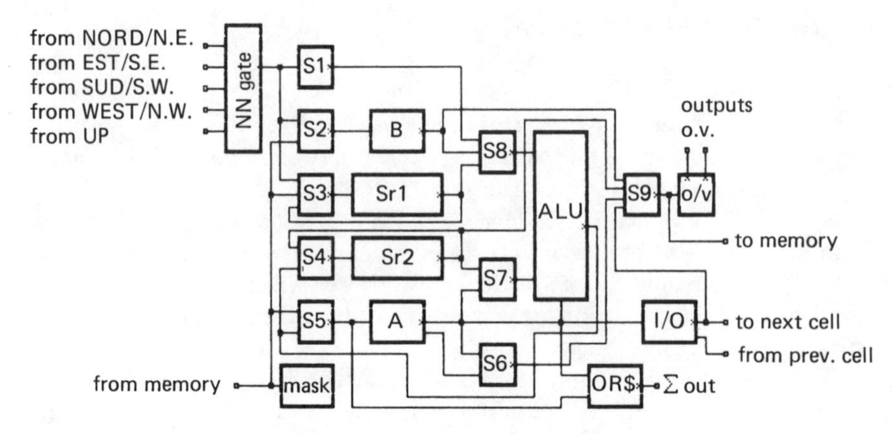

Fig. 2 Processor block diagram: NN gate interconnection logic; S1, S2, . . . S7, switch blocks; S8, switch and boolean block; S9, switch block

nonadjacent processors. In our project a second, faster, clock has been introduced to drive data along the selected path, implementing a distributed shift register in which each one of its cells corresponds to a specific register in the processor architecture.

In sequential data processing, the input/output bottleneck has been widely discussed; and different approaches, including partial overlapping with the processing, are widely used. In our case, the image input/output can be completely overlapped with the processing. The two previously mentioned clocks synchronize processing and input/output, respectively. The second, external clock can be run at a speed that depends on the specific input/output device. The pyramid layers are arrays; so the image data will be loaded column-wise at each cycle of the second clock. In this way n (number of columns) clock cycles plus one extra cycle of the first clock are required to copy, in or from memory, the contents of all the I/O registers,(a full loading/unloading image operation).

Many basic applications of parallel image processing can be efficiently implemented in a SIMD architecture by using recursive computation [14] which is data driven and must terminate whenever a stable data configuration is obtained. For this reason a global test is required on this system; namely, the exclusive-OR between the past and present states of all processors builds up a general OR-tree that provides a boolean response on the stability of the image data configuration. Moreover, global tests on the image content (empty, full) can be achieved by copying the content of the status register (in each processor) into the OR-tree circuitry.

A typical feature of a SIMD architecture is that all processors operate in a tightly orchestrated way and are unidentifiable (i.e., they cannot be individually addressed). To enable the selection of a subset of processors (from the whole array) a masking register that is in correspondence with each processor has been provided.

As previously mentioned, the pyramid structure can operate in both the SIMD and the Multi-SIMD modes; in the first case a subset of the layers can be activated by means of a plane masking register. As may be seen, two masking levels are provided; a processor will be working when both its layer masking register and its own masking register are enabled.

Although this processor is essentially a one-bit machine using serial arithmetic, its design includes two variable-length registers to handle grey-level images easily, and therefore to avoid the cumbersome memory-to-processor transfers. These transfers occur because of the bit-plane representation of grey-level images and frequently constitute a heavy burden for the memory management system.

3. CHIP CONFIGURATION AND LAYOUT

Many approaches exist for the partial or full automation of the layout process of logic circuit designs. These approaches aim at the generation of a formal definition of an interface between the chip designer and the chip manufacturer, according to the Mead-Conway philosophy which emphasizes the difference between the design phase and the construction phase of the project. We have followed the standard cell approach [15] in which some basic cells are defined a priori and optimized for subsequent interconnection to obtain the chip functional units.

The final topology has been defined by using a graphic editor KIC [16], and Fig. 3 shows an example of the resulting layout, which contains the main register of the processor element, i.e., a master–slave flip-flop.

Note, however, that in cases in which the regularity of the circuitry allows greater modularity, a programmable logic array approach has been followed. As an example, see Fig. 4, in which a length decoder of the shift registers is illustrated.

Two memory types for each processor are employed: one, of static nature and random access, is organized in a 32 × 8 array of single-bit cells, and one, of dynamic nature and sequential access, is organized in two variable-length (1, 2, 4, 8, 16, 32) shift registers. The basic cell of the static memory, made by six transistors, is shown in Fig. 5(a) and its corresponding stacked layout in Fig. 5(b). The basic dynamic shift register cell is shown in Fig. 6(a) and its layout in Fig. 6(b).

Let us now describe the full chip configuration (refer to Fig. 7). The top row shows the five processors; the central one is the father of the other four

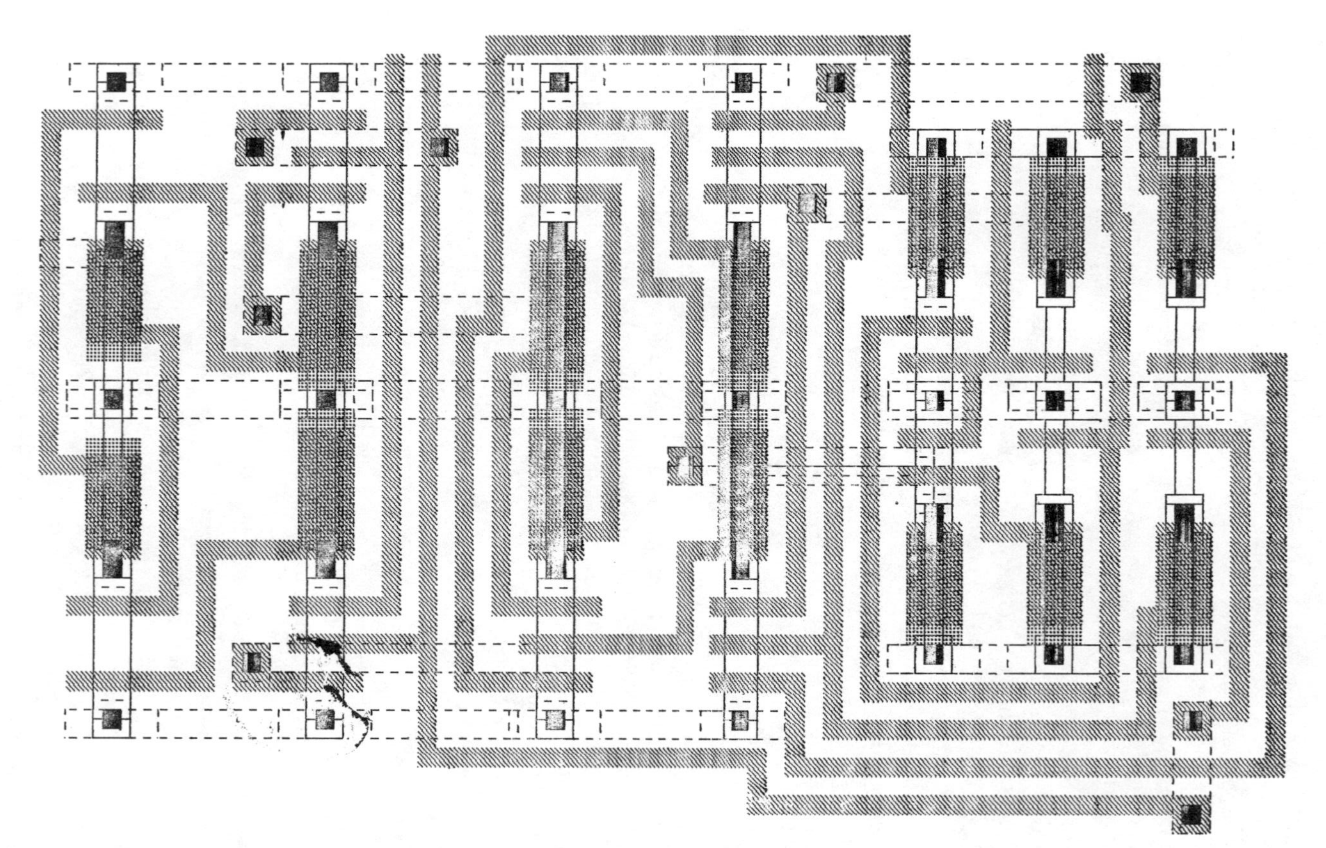

Fig. 3 Main register layout (master-slave flip-flop) by the standard cell approach (polysilicon is shown by diagonal shading, metal by dashed lines, and diffusion by solid lines)

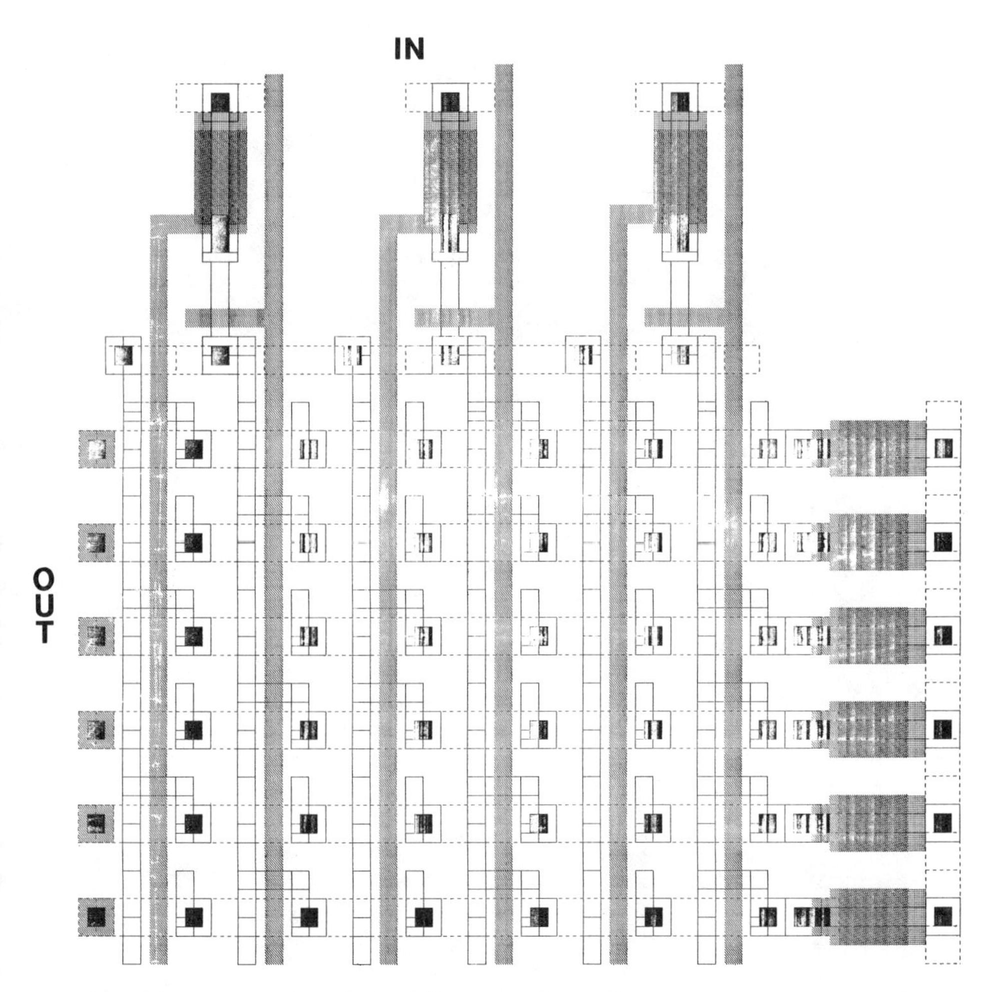

Fig. 4 Length decoder of the shift register layout by an approach similar to the programmable logic array

(sons) and should be considered as the elementary pyramidal unit of the system. On the right-hand side of each processor, a gating logic is provided to enable the reception of the inputs either from the four coplanar neighbors (brothers) or from the vertical neighbors, one from the upper layer (father) and four from the lower layer (sons). The second row of boxes shows ten shift registers (two for each processor). The physical length of each register has

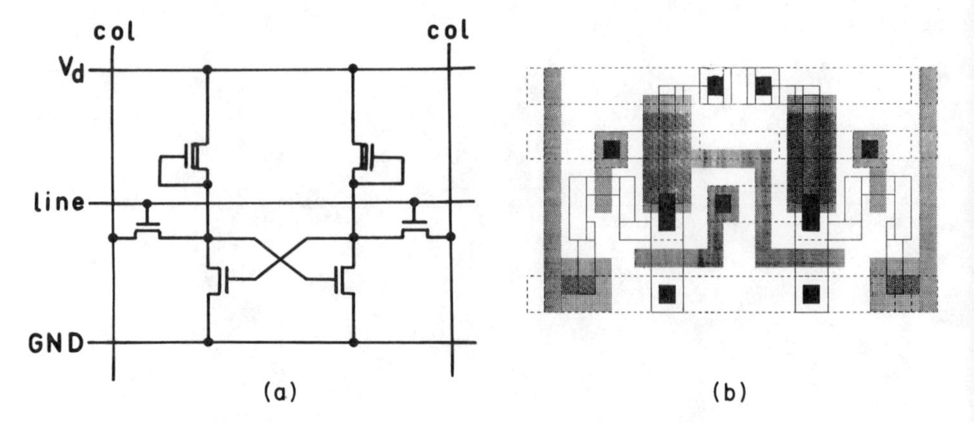

Fig. 5 The basic cell of the static memory: (a) electric diagram and (b) corresponding layout

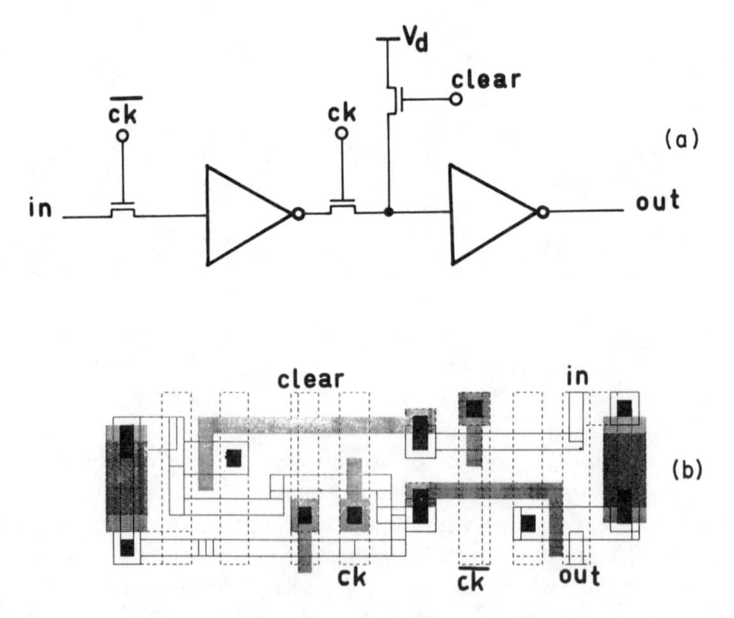

Fig. 6 The basic dynamic shift register cell: (a) electric diagram and (b) corresponding layout

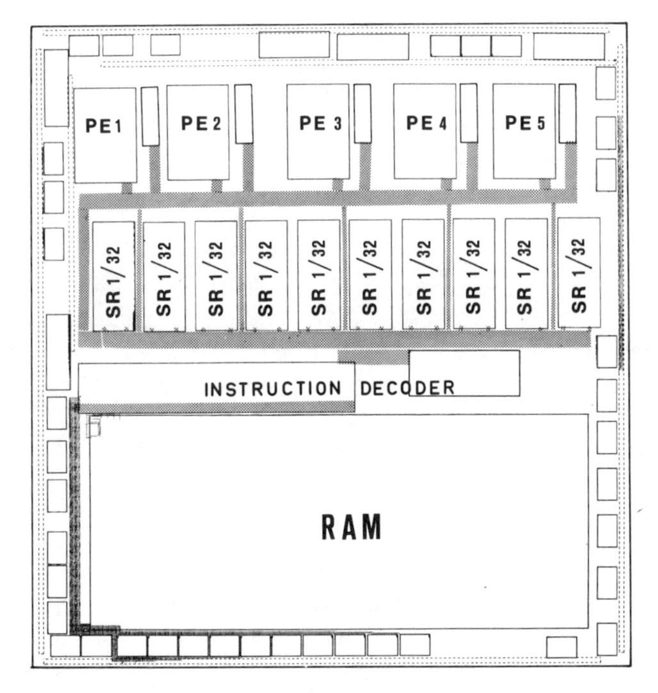

Fig. 7 The complete chip layout diagram

been halved by folding it along the y-axis. To reduce the interconnection tracks, each pair of registers is placed in mirror symmetry around the y-axis.

The next row of boxes shows the instruction decoder split into two parts: the first referring to the memory and the second to the processors. The instruction word-length is 12 bits for memory reference instructions, 8 bits for the address (the less significant bits), and the remaining four for the operational code; the last of these four bits indicates a memory reference instruction. This implies that only eight memory reference instructions can be defined: four for loading (into A and B registers, Sr1, and the mask register) and four for storing (from the A and B registers, Sr2, and the I/O register). Last, the bottom row contains the local RAM memory $32 \times 8 \times 5$ bits.

Pads are distributed all around the chip frame and lead to 48 pins of a dual-in-line package. The elementary pyramid has a neighborhood of 29 elements: 16 plus 1 vertical neighbors and 8 plus 4 horizontal ones. Five outputs must be communicated to the neighbors so that a total of 34 unidirectional lines are required if concurrent communications are to be ensured. Unfortunately, the pin number was limited to 48; so if all the other requirements were satisfied (clocks, supplies, instruction code, I/O, and OR-sum status of the chip), the available number of pins would have been exceeded. For this reason and because algorithms rarely require simultaneous horizontal and vertical communication, we have decided to share the input communication pins among horizontal and vertical neighborhoods. Note that with this solution, whenever required, two successive clock cycles ensure the access to the complete neighborhood. The 17 vertical inputs (shared with the 12 horizontal ones) must be incremented by the outputs, which must be doubled using a three-state logic to feed the horizontal or vertical channels.

This chip, currently being manufactured using a 4-μm Si-gate NMOS technology, has an area of 25 mm^2. The size of the basic cell, shown on Fig. 5, is of $98 \times 58\ \mu m^2$; and the whole RAM, including the address decoder (for rows and columns), occupies $4160 \times 2800\ \mu m^2$.

Standard cells are interconnected on two levels: the polysilicon level shown in Fig. 3, 4, 5(b), and 6(b) as diagonally shaded arc and the metal level outlined by dashed lines.

Each standard cell has been electrically simulated by means of the SPICE [17] software, and the propagation delay has been determined. The results can be exemplified by a simple inverter biased with 5 V and with a 500 fF load that produced a 5-ns delay time. Considering the introduction of a built-in body bias generator and the figures given above, we expect the maximum clock frequency to be about 5 MHz.

4. CONCLUSIONS

This chip will be available by the winter of 1984; and after the testing phase is completed, a new edition will be scheduled by the end of this year. In the meantime, the layer controller is being designed, and the host system is being evaluated and purchased. Referring to the internal organization of the machine, 17 chips will be assembled on each single board to implement an elementary pyramid with four layers with a total number of 257 boards.

To make the system usable by a large community of researchers, a high-level language must be provided. This language should contain instructions for expressing concurrency and suitable data types for the image data. Many versions of parallel Pascal are under study in different parts of the world [18], and we hope to learn from the experience accumulated in other groups [19].

The application areas of such a system range from low-level vision to graph algorithms and matrix computation.

ACKNOWLEDGEMENTS

The authors wish to thank Professor R. Stefanelli for useful discussions on serial arithmetic and C. Canobbio, L. Cinque, G. Varasano for their help in various phases of this work.

REFERENCES

[1] Cantoni, V., and Levialdi, S. (1983). Matching the task to an image processing architecture. *Computer Vision, Graphics and Image Processing*, **22**, 301–309.

[2] Tanimoto, S. L., and Klinger, A. (eds.). (1980). *Structured Computer Vision: Machine Perception through Hierarchical Computation Structures*. Academic Press, New York.

[3] Uhr, L. (1983). Pyramid multicomputer structures, and augmented pyramids. In *Computing Structures for Image Processing*. (M. J. B. Duff, ed.) Academic Press, New York, pp. 95–112.

[4] Dyer, C. R. (1982). Pyramid algorithms and machines. In *Multicomputers and Image Processing*. (K. Preston, Jr. and L. Uhr, eds.). Academic Press, New York, pp. 409–420.

[5] Samet, H. (1980). Region representation: quadtrees from binary array. *Computer Graphics and Image Processing*, **13**, 88–93.

[6] Freeman, H. (in press). Octree: a data structure for solid object modelling. In *Computer Architectures for Spatially Distributed Data*. (H. Freeman and G. Pieroni, eds.). Springer-Verlag, Berlin and New York.

[7] Rosenfeld, A. (1984). Multiresolution image representation. In *Digital Image Analysis* (S. Levialdi, ed.). Pitman, London, pp. 18–28.

[8] Kelly, M. D. (1971). Edge detection in pictures by computer using planning. *Machine intelligence 6*, Edinburgh University Press, Edinburgh pp. 397–409.

[9] Duff, M. J. B. (1976). CLIP4: A large scale integrated circuit array parallel processor. *Proc. 3rd IJCPR*, Coronado, pp. 728–733.

[10] Batcher, K. E. (1980). Design of a massively parallel processor. *IEEE Trans. Comput.* **C29**, 836–840.

[11] Tanimoto. S. L., and Pfeiffer, Jr., J. J. (1981). An image processor based on an array of pipelines. *IEEE Computer Society Workshop on Computer architecture for Pattern analysis and Image Database Management* Hot Springs, pp. 201–208.

[12] Reddaway, S. F. (1973). DAP: a distributed processor array. *First Annual Symposium on Computer Architecture*, Florida, pp. 61–65.

[13] Cantoni, V., Guerra, C., and Levialdi, S. (1983). Towards an evaluation of an image processing system. In *Computing Structures for Image Processing* (M. J. B. Duff, ed.). Academic Press, New York, pp. 43–56.

[14] Preston, Jr., K., Duff, M. J. B., Levialdi, S., Norgren, P. E., and Toriwaki, J. I. (1979). Basics of cellular logic with some applications in medical image processing. *Proc. IEEE*, **67**, no. 5, 826–856.

[15] Mead, C., and Conway, L. (1980). *Introduction to VLSI Systems*. Addison-Wesley, Reading Massachusetts.

[16] Keller, K., and Newton, A. R. (1982). KIC II: A low cost interactive editor for integrated circuit design. *IEEE Comcon 82*, San Francisco, pp. 302–304.

[17] Nagel, L. W. (1975). SPICE II: A computer program to simulate semiconductor circuits. ERL Memo, no. ERL-M520, University of California Press, Berkeley.

[18] Duff M. J. B., and S. Levialdi, (1981). (eds.) *Languages and Architectures for Image Processing*. Academic Press, London.

[19] Session on Image at the IEEE Workshop on *Languages for Automation*. Chicago, (1983.) *Computer Soc.*, no. 506, pp. 59–86.

Chapter Nine

*The Use and Design of PASM**

James T. Kuehn, Howard Jay Siegel,
David Lee Tuomenoksa,
and George B. Adams III

1. INTRODUCTION

Parallel processing has been successfully used to reduce the time of computation for a wide variety of applications. The processing of large amounts of data, the need for real-time computation, the use of computationally expensive operations, and other demands that would make a task too time-consuming to perform on conventional computer systems have forced computer architects to consider parallel/distributed computer designs. Applications that have one or more of these characteristic demands include image analysis for automated photo reconnaissance, map generation, robot (machine) vision, and rocket and missile tracking; digital signal processing for speech understanding and biomedical signal analysis; and vector processing for the solving of large systems of equations. To date, a variety of special-purpose machines has been constructed to speed the processing of select groups of algorithms. Examples are special-purpose digital signal processors such as the APS-II [1], array processors such as the AP-120B (Floating Point Systems, Inc. Portland, Oregon), and supercomputers with vector/pipeline operations such as the Cyber 205 [2].

Our goal is the design of a flexible parallel processing system that can be dynamically reconfigured to meet the particular processing needs of a large variety of applications in the image and speech analysis domains. The system

* The research was supported by the United States Army Research Office, Department of the Army, under grant number DAAG29-82-K-0101; by the United States Air Force Command, Rome Air Development Centre, under contract number F30602-83-K-0119; and by the National Science Foundation under grant ECS-81-20896.

INTEGRATED TECHNOLOGY FOR
PARALLEL IMAGE PROCESSING
ISBN 0-12-444820-8

being designed is a PASM, *pa*rtitionable *S*IMD/*M*IMD machine. In this chapter, two algorithms used in parallel contour extraction are given as an image processing scenario to explore the advantages and implications of using the PASM parallel processing system and to motivate the inclusion of its important architectural features. These features will help to identify the attributes of a custom-designed VLSI processor chip set for PASM. In particular, the architectural features that could be incorporated into a VLSI chip set that will match the needs of parallel algorithms in the image and speech processing domains will be explored. Using algorithm characteristics to drive the design of PASM will lead to a machine that has the necessary flexibility for executing image and speech processing algorithms.

In the next section, the parallel processing model and an overview of the PASM architecture are given. Section 3 outlines two algorithms of the contour-extraction task. The first algorithm, edge-guided thresholding, is discussed in Section 4. Section 5 describes the second algorithm, contour tracing. The architectural implications of these algorithms are explored in Section 6.

2. SIMD/MIMD MODEL

Two types of parallel processing systems are single-instruction stream–multiple-data stream (SIMD) machines and multiple-instruction stream–multiple-data stream (MIMD) machines [3]. A *SIMD machine* typically consists of a control unit, an interconnection network, and N *processing elements* (PEs), with each PE being a processor/memory pair (Fig. 1). The

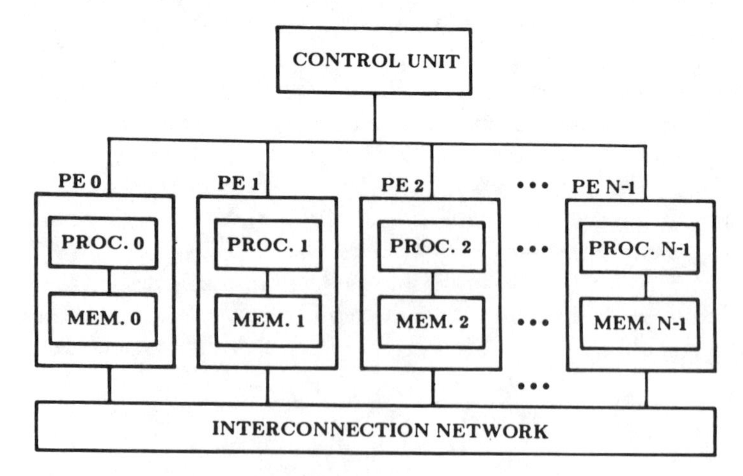

Fig. 1 Model of an SIMD/MIMD machine.

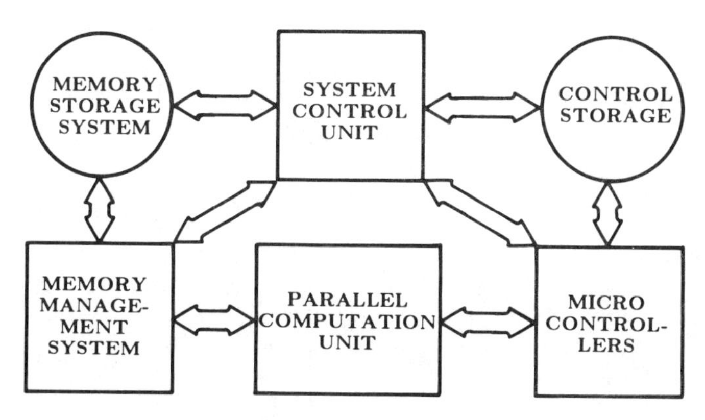

Fig. 2 Block diagram overview of PASM.

control unit broadcasts instructions to the processors, and all active (enabled) processors execute the same instruction at the same time. Each processor executes the instructions with data taken from its own memory. The interconnection network allows interprocessor communication. A *MIMD machine* has a similar organization, but each processor can follow an independent instruction stream. As with SIMD architectures, there is a multiple data stream and an interconnection network. The control unit may coordinate the activities of the PEs in MIMD mode. A SIMD/MIMD machine can operate in either mode and dynamically switch between them. A partitionable SIMD/MIMD system (e.g., PASM [4]; TRAC [5],[6]) can be dynamically reconfigured to operate as one or more independent SIMD/MIMD machines of various sizes.

PASM is being designed using a variety of applications problems from the areas of image and speech analysis to guide the machine design choices. It is not meant to be a production-line machine but a research tool for studying large-scale SIMD and MIMD parallelism.

A block diagram of the basic components of PASM is given in Fig. 2. The heart of the system is the *parallel computation unit* (PCU), which contains $N = 2^n$ processors, N memory modules, and an interconnection network. The PCU *processors* are microprocessors that perform the SIMD and MIMD computations. The PCU *memory modules* are used by the PCU processors for data storage in SIMD mode and both data and instruction storage in MIMD mode. The *interconnection network* provides communication among the PEs. PASM will use either an Extra Stage Cube type or Augmented Data Manipulator type of multistage network [7].

The PCU is organized as shown in Fig. 3. Each processor is connected to a memory module to form a PE. A pair of memory units is used for each

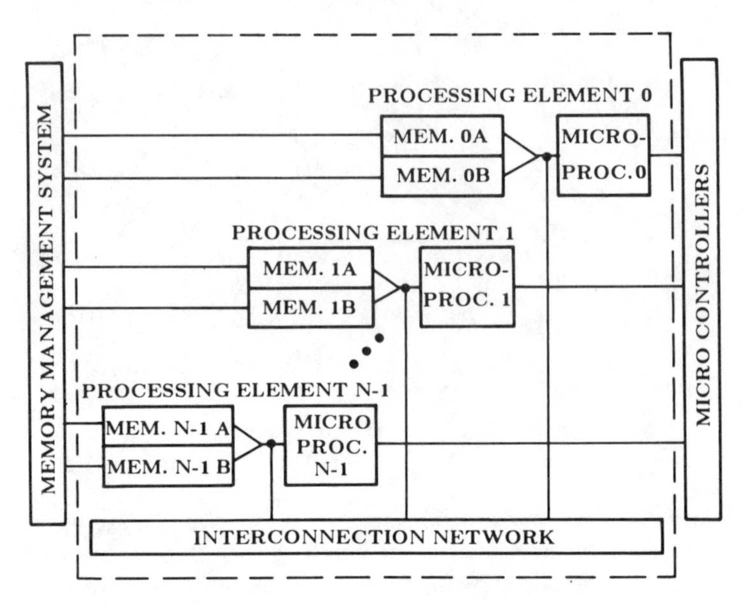

Fig. 3 PASM Parallel Computation Unit.

memory module. This double-buffering scheme allows data to be moved between one memory unit and secondary storage while the processor operates on data in the other memory unit. Each memory unit is of substantial size (e.g., 64K words). PEs are addressed (numbered) from 0 to $N - 1$.

The *system control unit*, a conventional computer, is responsible for the overall coordination of the activities of the other components of PASM. The *memory management system* controls the loading and unloading of the PE memory modules from the multiple secondary storage devices of the *memory storage system*. The *microcontrollers* (MCs) are a set of microprocessors that act as the control units for the PEs in SIMD mode and orchestrate the activities of the PEs in MIMD mode. Each of the Q MCs controls a fixed group of N/Q PCU PEs. By combining the effects of multiple MCs, virtual machines (partitions) can be created. *Control storage* contains the programs for the MCs. PASM is being designed for $N = 1024$ and $Q = 32$. An $N = 16$, $Q = 4$ prototype based on Motorola MC68000 processors is under development [8].

This brief overview of PASM provides the needed background for this chapter. Further details and a list of papers about PASM can be found in Siegel [9].

3. EXAMPLE TASK

Many individual image and speech processing algorithms and their formulations for parallel processing environments have been studied, such as 2-D FFTs [10],[11], Hadamard transforms [12], image correlation [13], histogramming [4], resampling [14], one-dimensional FFTs [15], linear predictive coding [16], and dynamic time warping [17]. However, rarely is a complete scenario considered as a whole. Consider the situation in which the results of one algorithm are used as input to another. In the parallel environment, this may strongly influence how each algorithm is structured. For example, results calculated in one PE might need to be communicated to another PE for use in a later algorithm.

Contour extraction is a key tool for use in applications ranging from computer assisted cartography to industrial inspection. Two algorithms from a contour extraction task will be used as an application example for demonstrating the architectural features that must be provided by PASM to have an appropriate execution environment. It will be shown how computational attributes of a parallel implementation of this example SIMD/MIMD scenario influence the hardware design choices, including those features that would be desirable in a custom-designed VLSI chip set.

The two algorithms to be considered are *edge-guided thresholding* (EGT) and *contour tracing*. The EGT algorithm, discussed in Section 4, is used to determine the optimal threshold for quantizing the image [18]. The contour-tracing algorithm, considered in Section 5, uses the set of optimal thresholds to segment the image and trace the contours. These two parallel algorithms are based on those developed in Tuomenoksa *et al.* [19] and are summarized here because their processing demands are quite different from each other. As will be seen, the EGT algorithm is best suited for SIMD mode, whereas MIMD mode will be used for the contour tracing algorithm. Also, the EGT algorithm will have inter-PE communication needs that are different from the communication needs of the contour tracing algorithm. Other aspects will be discussed in Section 6. For this task scenario, the ability to partition PASM is not used; i.e., all N PEs are employed.

4. EDGE-GUIDED THRESHOLDING

Consider an $M \times M$ pixel *input image* to be processed by the two algorithms. The value of each pixel is assumed to be an 8-bit unsigned integer representing one of 256 possible gray levels. Using the PASM model, assume that the PEs are logically configured as a $\sqrt{N} \times \sqrt{N}$ grid, on which the $M \times M$ image is superimposed; i.e., each processor has an

$M/\sqrt{N} \times M/\sqrt{N}$ *subimage*. For $M = 4096$, $N = 1024$, each PE stores a 128×128 subimage. Each input image pixel is uniquely addressed by its i–x–y coordinates, where x and y are the x–y coordinates of the pixel in the subimage contained in PE i.

The EGT algorithm consists of three major steps. First, the Sobel edge operator [20] is used to generate an *edge image* in which gray levels indicate the magnitude of the gradient. A figure of merit that indicates how well a given thresholded gray-level image matches edges in the edge image is then computed for every possible threshold. Finally, the maximum value of the figure-of-merit function is chosen to determine the threshold level. This is done for each subimage independently; thus, the threshold levels may differ from one subimage to the next. The complete EGT algorithm is most easily formulated as the SIMD procedure given in Fig. 4. Let the subimage SI be $M/\sqrt{N} \times M/\sqrt{N}$ and $\mathrm{SI}(i,x,y)$ be a subimage pixel, where $0 \leq x, y < M/\sqrt{N}$, $0 \leq i < N$. The algorithm is performed for all of the subimages (all i) simultaneously.

Referring to Fig. 4, the first *for* statement clears the sumedge and nedge counters (to be described) for each possible threshold value. The next pair of nested *for* statements contains statements to calculate quantities associated with each pixel in the subimage. The Sobel operators, *sx* and *sy*, represent weighted pixel value differences in the x and y directions, respectively. The value $g(i,x,y)$ represents the gradient at pixel (i,x,y), and these values form the edge image. The presence of an edge is indicated by high edge image pixel values. Next, the local maximum and minimum pixel values over a 3×3 window are determined for each gray-level image pixel. Note that the same image pixels necessary for the calculation of the gradient can be re-used for the determination of the local maximum and minimum. The center pixel of the 3×3 window is an *edge point* if the threshold is greater than or equal to the local minimum and less than the local maximum. Running sums of the edge image pixels (gradient values) corresponding to edge points at each threshold (sumedge) and a count of the number of edge pixels for each threshold (nedge) are updated in the innermost *for* loop. In general, each PE performs this *for* statement using a different localmin and localmax and thus performs the statements in the loop (updates the sums) various numbers of times. This implies that each PE has the capability of maintaining its own loop index values. PEs are disabled when they finish their looping, because PEs must remain synchronized in SIMD mode. The total time to perform the innermost *for* loop is the maximum time taken by any PE.

The mean for each threshold (sumedge/nedge) is known as the figure of merit (merit) and is calculated in the final *for* statement using the accumulated sums. High figure of merit values indicate better matches between threshold-generated boundaries and the edges detected by the Sobel oper-

```
for thresh = 0 to 255 do
   sumedge(i, thresh) = nedge(i, thresh) = 0
for x = 0 to M/√N − 1 do begin
   for y = 0 to M/√N − 1 do begin
      sx(i,x,y) = ¼[(SI(i,x − 1,y − 1) + 2*SI(i,x − 1,y)
                  + SI(i,x − 1,y + 1)) − (SI(i,x + 1,y − 1)
                  + 2*SI(i,x + 1,y) + SI(i,x + 1,y + 1))]
      sy(i,x,y) = ¼[(SI(i,x − 1,y − 1) + 2*SI(i,x,y − 1)
                  + SI(i,x + 1,y − 1)) − (SI(i,x − 1,y + 1)
                  + 2*SI(i,x,y + 1) + SI(i,x + 1,y + 1))]
      g(i,x,y) = √(sx(i,x,y)² + sy(i,x,y)²)
      localmax(i)
         = max(SI(i,x − 1,y − 1), SI(i,x,y − 1), SI(i,x + 1,y − 1),
               SI(i,x − 1,y), SI(i,x,y), SI(i,x + 1,y),
               SI(i,x − 1,y + 1), SI(i,x,y + 1), SI(i,x + 1,y + 1))
      localmin(i)
         = min(SI(i,x − 1,y − 1), SI(i,x,y − 1), SI(i,x + 1,y − 1),
               SI(i,x − 1,y), SI(i,x,y), SI(i,x + 1,y),
               SI(i,x − 1,y + 1), SI(i,x,y + 1), SI(i,x + 1,y + 1))
      for thresh = localmin(i) to localmax(i) − 1 do begin
         sumedge(i, thresh) = sumedge(i, thresh) + g(i,x,y)
         nedge(i, thresh) = nedge(i, thresh) + 1
         end
   end
for thresh = 0 to 255 do
   merit(i, thresh) = sumedge(i, thresh)/nedge(i, thresh)
```

Fig. 4 EGT algorithm.

ator. The gray level associated with the maximum value of the figure-of-merit function is chosen for image segmentation.

The EGT algorithm is particularly well suited for SIMD parallelism because all pixels are processed similarly. This aids the PE-to-PE communication necessary when a PE must process pixels not in its subimage (i.e., in a neighbor PE). All PEs will simultaneously request the same pixel relative to their subimages. For example, when processing begins (with the upper left corner subimage pixel) all PEs will request (from the PE to their upper left) the pixel immediately above and to the left of their upper left corner pixel (if this pixel is in the complete image). This transfer of data from upper left neighbors can occur for all PEs simultaneously. In the case of this algorithm, transmission delays incurred due to PE-to-PE data transfers can be overlapped with data processing to reduce total execution time. A total of $4(M/\sqrt{N} + 1)$ parallel transfers are needed for a $M \times M$ pixel image. The candidate interconnection networks of PASM can support these parallel transfers from any neighboring PE.

Since PE-to-PE communications in MIMD mode require explicit synchronization between the two processors for each data transfer, SIMD mode transfers should be used to provide each PE more efficiently with the one-pixel-deep border points of its subimage (from its neighbors). However, once each PE has all of the data it needs to perform the EGT algorithm, the calculations could proceed in MIMD mode. Although MIMD mode would make the execution of the innermost *for* loop more efficient (because no PEs would be disabled), this advantage must be weighed against the extra time involved in switching from SIMD to MIMD mode and requiring that each PE perform its own control flow operations for the outer two *for* loops. Control flow operations include initialization and incrementing of loop counters, evaluation of conditional expressions, and branching. These operations are performed by the MC in SIMD mode for the outer two loops and can be overlapped with the PE operations. The next step of the scenario is contour tracing.

5. CONTOUR TRACING

A contour tracing algorithm using MIMD parallelism and based on the one given in Tuomenoksa *et al.* [19] is summarized in this section. Initially, each PE contains a threshold value T for its subimage, which was calculated using the EGT algorithm of the previous section. The contour tracing algorithm has two phases. In Phase I, the PEs segment their subimages based on the threshold and all local contours (both closed and partial) are traced and recorded. In Phase II, the partial contours traced during Phase I are connected.

A *contour table* is constructed in each PE, containing an entry for every partial or complete contour in its subimage. Each contour table entry contains bookkeeping information such as the threshold value that generated the contour and a pointer to the the i–x–y sequence of the contour. Each PE also contains a *partial contour list*, which has an entry for each partial contour containing the i–x–y coordinates of its two end points and a pointer to its contour table entry.

In Phase I there is no PE-to-PE communication. Each PE uses its threshold level to segment its subimage. To create the segmented image for threshold T, subimage pixels that have a value greater than or equal to T are assigned a value of one; otherwise, the pixels are assigned a value of zero.

Contour tracing begins by scanning rows of the segmented image beginning with the top row. Scanning stops when a pixel with a value of one is found that has a zero-valued neighbor on both sides. This pixel is marked as the *start point* of a new contour, and its i–x–y coordinates are stored. For edge PEs, i.e., those on the edge of the $\sqrt{N} \times \sqrt{N}$ grid of PEs, no image points lie beyond the edge; thus, all points in the leftmost (or rightmost) column of the subimage of the PEs in the leftmost (or rightmost) column of the grid of PEs are potential start points. For all other left and right subimage edges, it is assumed that the pixel in the neighboring PE is one-valued so that spurious start points are not chosen. Bypassing a potential start point (e.g., a left subimage edge with a zero-valued neighbor in the PE to its left) is not a problem because (1) contours have multiple potential start points within the subimage and (2) the partial contours will be connected in Phase II regardless of the start point chosen.

The contour is first traced in a counterclockwise direction (CCW) if the start point has a one-valued point to its right and is first traced in a clockwise direction (CW) if the start point has a one-valued point to its left. If there are zeroes on both sides, the initial direction chosen does not matter. Consider the start point pixel as the center pixel of the 3×3 window in which direction 0 is east, 1 is northeast, and so on [21]. The CCW algorithm is stated as follows. Beginning with the neighboring pixel in direction five and incrementing by 1 modulo 8 to determine the next pixel, look for a pixel that has a value of one. When it is found, store the direction p of this new pixel and append its i–x–y coordinate to the contour sequence. Treat this pixel as a new center point of the 3×3 window. Then continue by looking for the next pixel in the contour beginning with the pixel in position $(p + 5)$ modulo 8. Tracing continues until the start point or a subimage boundary (point of indecision) is reached. The CW algorithm is similar, but scanning begins with the pixel in position zero and decrements by 1 modulo 8 to determine the next pixel. After a point is found, the pixel in position $(p + 3)$ modulo 8 is scanned. Horizontal edges that span a subimage are also recognized; however, they are treated as a special case because no start point would have been identified. An implicit assumption is that all contours to be traced define regions that have area. Examples of illegal contours that would not be traced are one-pixel-wide lines or isolated points.

A *point of indecision* is reached when a pixel from an adjacent subimage is needed to determine the next direction of the contour [19]. When a point of indecision is reached, it is recorded as an *end point*, and the algorithm returns to the start point to trace the contour in the opposite direction until another point of indecision is reached. When tracing in the CW (or CCW) direction, the new contour pixels are inserted onto the front (or back) of the i–x–y sequence. Pixels in the thresholded image are marked so that the contour will not be retraced.

As an example, a 30 × 20 image is divided into six 10 × 10 subimages; each subimage is loaded into one of six PEs. The result of Phase I processing is shown in Fig. 5 where a dot indicates a one-valued pixel. Even though the entire object in PE 5 was located within the subimage, the left edge of the object was not traced in Phase I, because PE 5 could not determine whether the object continued into the next subimage. On the other hand, a closed contour was found in Phase I for the object in PE 4, because the object did not include any border pixels of the subimage.

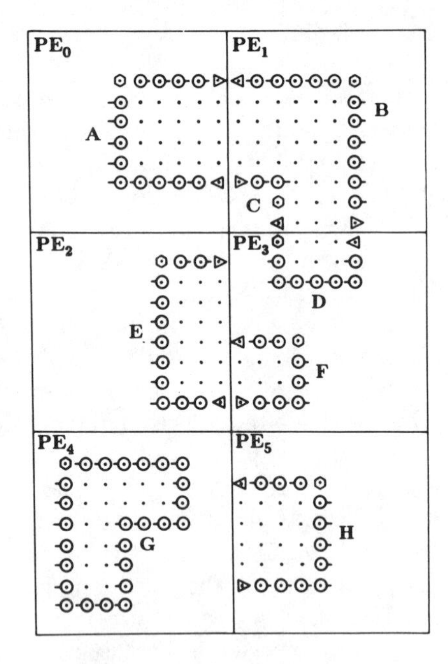

Fig. 5 Results of Phase I of contour tracing for a 30 × 20 subimage. (Based on Tuomenoksa, *et al.*, [19].)

For the example in Fig. 5, the local threshold value T is applied to the subimage in each PE. Each PE i begins scanning its respective subimage at pixel (i, 0, 0,) for a one (indicated by a dot) with a zero on either side. Depending on the start point found, tracing will proceed in either the CW or CCW direction. For example, contours A, C, E, D, and G are traced in the CCW direction first, whereas contours B, F, and H are traced in the CW direction first. In the example, PEs 1 and 3 have found two start points and have produced two traces. Once a PE has scanned the segmented image generated by its threshold, Phase I is complete.

In Phase II, each PE attempts to connect its partial contours to those located in neighboring PEs. In order for a PE to extend a contour, it must be able to access and modify contour tables that are located in other PEs. As a result, a mechanism to allow access to a contour table entry by only one PE at a time must be provided by the system and used by the contour tracing algorithm. A *semaphore* [22] associated with each contour table entry is used to indicate whether or not that entry is locked so that no other processor can access it. Semaphores are used to prevent variable access and updating problems due to interrupts. Details of these problems are beyond the scope of this paper.

For the example of Fig. 5, PE 0 might try to extend the CW end point of partial contour A by considering the possible extending pixels in PE 1 one at a time using the CW algorithm. To do this, PE 0 first locks the contour table entry for A. Then PE 0 requests that PE 1 check its partial contour lists to determine if any partial contour has the possible extending point as an end point. If such a partial contour exists, PE 1 locks the contour table entry pointed to by the partial contour list signifying that this entry is to be linked. In this case, PE 0 determines that A can be linked to B; thus, PE 1 locks B's contour table entry so that only PE 0 will be allowed to connect the partial contour. The i–x–y sequence for contour B is transferred to PE 0 and concatenated to the i–x–y sequence of partial contour A, forming a new, extended partial contour AB. If PE 0 found the contour table of partial contour B to be already locked, it will not be allowed to connect the contour. The extension of corner points is handled similarly but involves communication with more than one PE. Note that the use of semaphores prevents another PE, i.e., PE 3, from using PE 1 to access B's contour table entry which PE 1 is in the process of modifying for PE 0.

Once PE i locates a partial contour in an adjacent subimage that continues the contour and has stored the concatenated contour in its contour table, it repeats the process, if necessary, by following the contour to the next PE until the contour is closed or cannot be extended.

Independently of the actions of PE 0, PE 3 might attempt to extend contour D CCW to form the partial contour DC. If PE 3 attempted to extend

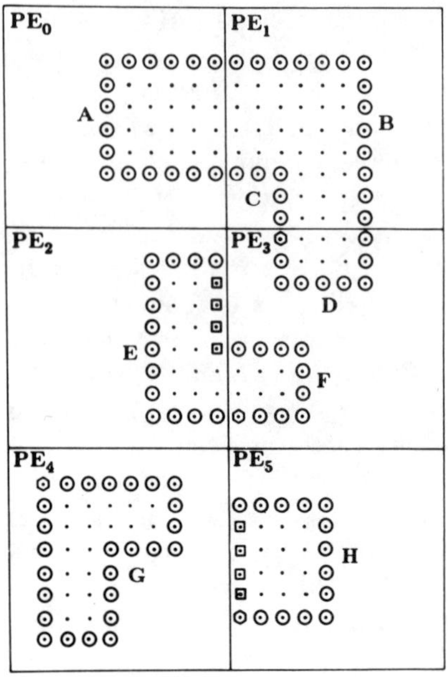

⊙ Pixels traced in Phase I
⊡ Pixels traced in Phase II
⊙ First pixel in the x-y sequence of the contour

Fig. 6 Results of Phase II of contour tracing for a 30 × 20 subimage. (Based on Tuomenoksa, *et al.*, [19].)

the result, DC, when PE 0 is in the process of extending A into PE 1, it will find A locked. PE 3 then abandons its attempt to close the contour, because PE 0 is also attempting to do it, and unlocks partial contour DC. This allows PE 0 to access DC after it has appended B to A. Therefore, the closed contour ABDC is ultimately traced by PE 0. Alternatively, if PE 0 had completed linking B to A before PE 3 completed linking C to D, and PE 0 finds D locked, it would unlock AB. Thus, the closed contour would have been completely traced by PE 3. Not allowing a PE to wait for access to another PEs locked contour table entry and requiring the blocked PE to unlock its affected partial contour prevents deadlock.

Occasionally, some contour tracing operations must be performed in Phase II before certain contours can be linked. Figure 6 shows a situation in which PE 2 traces contour E along the subimage boundary in Phase II before linking it to contour F. The subimage boundary pixels of contour H are also traced in Phase II.

These examples demonstrate the basic ideas underlying the algorithms. The actual parallel algorithm details that ensure proper interaction of the PEs are complex and are not examined here.

When Phase II of the algorithm is complete, the i–x–y sequence for each contour in the image will be contained in exactly one of the PEs that contained part of the contour originally. The result of Phase II processing for the example for Fig. 5 is shown in Fig. 6. Since each PE tries to connect its contours independently, the number of the PE that finally closes a given contour is nondeterministic. Although this may not be desirable in a few cases, in general the lack of a specific protocol determining which PEs can close contours equalizes both the processing load of each PE and the number of closed contours that eventually reside in each PE.

6. ARCHITECTURAL IMPLICATIONS

The study of a parallel image processing task leads to an understanding of necessary and useful hardware features for a system such as PASM. For the example algorithms, aspects of each that have an architectural impact will be listed. Processor-specific considerations (e.g., instruction set) are also treated, because they can have a profound effect on the performance of the algorithms.

Although only two closely related algorithms were presented in the previous sections, the two could hardly have been more different in their processing demands. As discussed in Section 4, the EGT algorithm is best suited for SIMD mode. This is because the algorithm requires data that are mostly local to each PE. Also, there are approximately (or exactly) the same number of pixels to be processed in each PE, and all pixels are processed similarly.

When nonlocal data are needed in the EGT algorithm, the eight nearest-neighbor PEs comprise the set of data sources. The PE-to-PE transfer of information must be efficient, or the parallel algorithms will be slowed. In its simplest form, this communication would be handled entirely by the PEs; each PE would control the network settings (through the use of routing tags [7]) and perform all of the network protocol support (e.g., buffering, error detection). Each word transferred and each new network setting would require processor instructions. A more efficient method of PE-to-PE communication is by *direct memory access* (DMA). DMA is a means by which data can be retrieved from one memory location and stored in another without processor intervention. The DMA hardware usually operates on a cycle-stealing basis so that a PE's access to its memory is not severely affected. In its basic form, PEs in SIMD mode would enter a DMA handling routine.

This routine computes the local memory address range of the points to be transferred and sends this information to the special DMA hardware. The PEs then would compute the destination address of the PE that is to receive the data and set the network accordingly. The DMA hardware then would autonomously retrieve the information from local memory and perform the necessary network interfacing to send the data to the requesting PE. However, the PE would still be responsible for checking the incoming data (after the transfer is complete) for transmission errors and so forth. A more advanced implementation of DMA capability is the use of an intelligent *network interface unit* (NIU). Requests for data from remote PEs would be made to the local NIU, which would interpret and satisfy the request by coordinating with a remote NIU. The NIU would combine DMA capability with network protocol support. VLSI technology may allow ready fabrication of sophisticated NIUs.

As discussed in Section 4, $M/\sqrt{N}$ pixels come from each of four neighbors. For the sake of example, let $M = 4096$, $N = 1024$, and $M/\sqrt{N} = 128$. Rather than involving the *source* and *destination* PEs in the individual transfers of these points, one of the DMA modes just described would be of great use. If pixels were stored in PEs by row (rows numbered 0–127), and the transfer from PE i to PE $i + \sqrt{N}$ was selected, the DMA hardware of PE i would be instructed to transfer 128 pixels starting at the address of row 127 of the image. The DMA hardware associated with PE $i + \sqrt{N}$ would be set to read 128 pixels from the network and store them beginning at an address representing row -1. When data are transferred from a PE i to PE $i + 1$, the situation is more complicated in that image data to be transferred are not contiguous. Conventional DMA hardware only supports physical block transfers of data. Here, a strong case for an intelligient NIU is made: the NIU could accept more complicated instructions such as "transfer 128 pixels starting at address X, taking every 128th pixel."

The processing requirements (instruction set) for the EGT algorithm are not out of the ordinary. LSI technology already allows the fabrication of complete microprocessors having all required arithmetic and data manipulation operations on a single chip. Recent designs (e.g., Motorola 68000 [23]) handle a variety of data formats including bit, byte, 16-bit word, and 32-bit long word types. Floating point and special arithmetic function (e.g., square root, trigonometric) capability abounds in the form of *coprocessor* chips. Although the EGT algorithm involved only one special function (square root) in the calculation of the gradient, other algorithms such as image rotation, parallel root finding, and FFTs for speech processing make heavy use of special functions. Since many of the special arithmetic functions are calculated by iterative procedures, a strong case is made for including hardware to perform these operations rather than performing them in software. Software procedures in which the number of iterations required is data-dependent are especially troublesome in SIMD mode, because processors must be disabled as they complete the desired number of iterations. Also, the total time to perform an operation is the maximum time required by any processor (because the PEs are synchronized). There is a slight advantage to having special-purpose arithmetic functions on the same standard CPU chip in that data to be processed need not be moved between the two devices. VLSI technology should make such combined CPU-specialized arithmetic processor chips a reality.

The processors must be capable of operating in SIMD mode efficiently. Although designs for using off-the-shelf microprocessors as SIMD/MIMD processing elements have been developed, some external hardware would be required to enable, disable, and synchronize PEs and get them to operate in slave mode, i.e., to accept instructions broadcast by a control unit rather than to take the instructions from their local memories [8]. This external hardware could be easily incorporated into a VLSI chip.

The EGT algorithm has been simulated for N ranging from 16 to 256 and a total image size of 64×64 pixels. A special-purpose SIMD simulator developed to evaluate the MC68000-based PASM design described in Kuehn *et al.*, [8] was used to perform the simulations. Although the details of the simulation results are not presented here, the general trends of the results will be described.

As the number of PEs (N) decreased, the subimage size increased because a fixed-size total image of 64×64 pixels was used. For large subimages, the ratio of subimage edge pixels to total subimage pixels is low, making processing very efficient. This is because inter-PE transfers make up only a very small fraction of the total processing time. A *speedup factor* (serial execution time/parallel execution time) approaching N was obtained for arithmetic operations for this case. (A speedup of N is optimal.) As N was increased to 256 PEs, the subimage size decreased to 4×4. Here, the ratio of subimage edge pixels to total subimage pixels is very high, and inter-PE transfers make up a large percentage of the total processing time. Although the total processing time is minimized as N increases, the speedup factor decreases. The simulations imply that N should be as large as possible for the EGT algorithm to minimize the processing time. However, this would make contour tracing (the next algorithm of the scenario) inefficient, because few contours would be traced in Phase I, and heavy use of inter-PE communication would be needed to close the contours in Phase II. Thus, the scenario must be considered as a whole rather than as a sequence of individual algorithms.

Turning now to the contour tracing algorithm of Section 5 we note that both phases of the algorithm are suited to MIMD mode, because they involve data-dependent execution times. Phase I of contour tracing requires only local data, whereas Phase II makes heavy use of nonlocal data. Phase I imposes no extraordinary requirements on the system, because there are no special arithmetic operations and no network transfers to be done. Phase II, however, with its arbitrary one-to-one connections (when transferring partial contour information between nonadjacent PEs), use of semaphores, and special signaling protocols imposes many new architectural requirements.

The interconnection network and any DMA or NIU hardware would be heavily used in Phase II processing when PEs extending partial contours probe remote PE memories that may contain the extensions of the partial contours. As in the EGT algorithm, NIU hardware would be of great use, because it could process queries about possible extensions to partial contours without interrupting the remote PE. There would be a combination of short and long messages between PEs during this phase. A short message would occur when a PE, extending a partial contour, requests information about possible extending pixels from a remote PE. If a connecting partial contour is found, a long message, consisting of the *i–x–y* sequence of the partial contour, would be sent. Thus the interconnection network should support a variety of message sizes so that the efficiency of sending either type of message is high.

Since semaphores play a large part in ensuring correct linking of partial contours in Phase II, processors must be equipped with *test-and-set* or similar operations to facilitate a correct semaphore implementation. Most modern microprocessors already have some semaphore capabilities.

If the system is to support the execution of the two example algorithms well, it must be capable of dynamically switching between SIMD and MIMD operation, as PASM can. With only SIMD capability, the contour tracing algorithm would be executed with huge inefficiencies, because there would be varying numbers and lengths of contours and arbitrary one-to-one communication patterns. A machine having only MIMD mode would be less seriously affected but would lengthen execution time for the EGT algorithm, due to the need for explicit synchronization for each data transfer step and the overhead of loop counter processing which is done concurrently by the MCs in SIMD mode. Thus, the capability to dynamically switch between SIMD and MIMD modes is important so that each algorithm can be executed in the most appropriate mode of parallelism.

Since PASM is an SIMD/MIMD system, the interconnection networks proposed for PASM would be capable of operating both synchronously and asynchronously. The proposed networks are of the multistage type and can perform both the nearest-neighbor and arbitrary one-to-one connections.

The design of a multi-microprocessor system that could be used as a building block for PASM is discussed in Kuehn *et al.*, [8]. This design uses the Motorola MC68000 as the heart of both the PE and MC components. The extra hardware needed for SIMD/MIMD mode processing and communication was described. It was found that most of the extra hardware was involved in the enabling/disabling, synchronization, and instruction broadcasting for SIMD mode and in getting the PEs to switch from SIMD to MIMD mode and back again efficiently. The design highlights are described:

MC CPU. The MC CPU is a Motorola MC68000-series processor.

Fetch unit. This unit fetches instructions from MC memory in SIMD mode, determines whether they are control (MC) or data processing (PE) instructions, and broadcasts them either to the MC CPU or PE CPUs. Each instruction word in the MC memory is tagged to allow the fetch unit to determine its type. The tags are generated at assembly time.

Masking operations unit. This is specialized hardware, under the control of the MC CPU, that produces a *mask* (pattern) used to selectively enable or disable PEs (used in SIMD mode).

MC/PE interface. This is specialized hardware to queue PE instructions and enable signals broadcast to the PEs. The queue has been shown to increase the amount of program overlap between the MC and PEs. This interface is for SIMD mode; there would also be a MC/PE communication bus for MIMD mode and error-handling messages (which is not discussed here).

PE CPU. The PE CPU is a Motorola MC68000-series processor.

SIMD/MIMD mode switching logic. This is a specialized address decoder that generates instruction requests to the MC/PE interface in SIMD mode and causes local PE memory to be accessed in MIMD mode.

Network interface unit. This unit handles DMA and network protocol.

VLSI technology should be used to combine the components listed above only when some speed or complexity advantage is gained. For example, the PE CPU and SIMD/MIMD mode switching logic should be combined into a single component so that the PEs can operate equally well in SIMD and MIMD mode. This action would result in very little additional silicon area and at most a few additional pins being used. Taking this one step further, one could also fabricate the DMA and NIU hardware on the PE CPU chip. However, to allow communication on the CPU data bus (with, for example, the local memory chips) and the NIU-interconnection network bus to occur simultaneously, pins for a complete NIU bus interface would have to be added. The technology at implementation time would determine the maximum pin count and thus the suitability of this scheme.

Similarly, the MC CPU and fetch unit should be combined on one chip so that MC operations such as fetching SIMD instructions and branching are done by the same unit. The masking operations unit could easily be made a part of the MC CPU since it is not too complex; however, the number of CPU pins would have to increase by N/Q. For the PASM design goal of $N = 1024$ and $Q = 32$, $N/Q = 32$. Again, the desirability of integrating this unit is dependent on pin count limitations. The MC/PE interface is also a candidate for inclusion on the MC chip. It would not require much silicon area, but its pin requirements are high. Since the interface queues both enable signals and instruction words to be broadcast to the PEs, an additional $N/Q + 16$ bits would be required on the MC CPU package (for MC68000 16-bit words). Thus, assuming that the number of pins that the MC CPU alone requires is P, if the masking operations is integrated with the CPU, $P + N/Q$ pins would be required; if, in addition, the MC/PE interface is integrated, $P + N/Q + 16$ pins would be required (the masking operations unit output pins to the MC/PE interface would now be internal to the chip). As has been discussed in McMillen and Siegel [24], VLSI implementation of interconnection network functions is most promising, both from a functional standpoint and a design standpoint due to network regularity.

In summary, based on our prototype plans and the expected execution needs of the contour extraction task and other image and speech processing algorithms, certain desirable system architecture features have been identified. These include dynamically switchable SIMD/MIMD capability, support for PE-to-PE communications using DMA and intelligent network interfaces, and special arithmetic function hardware. These requirements are consistent with the capabilities of a VLSI implementation of PASM.

7. SUMMARY

Contour extraction has been used as an image processing scenario to explore the advantages and implications of using the PASM parallel processing system. Use of these parallel algorithms leads to several advantages, notably speedup. Analysis of the algorithms has motivated the inclusion of several important architectural features. These features were used to discuss possible configurations of a custom-designed VLSI processor chip set for PASM. The use of algorithm characteristics to drive the design of PASM leads to a machine with features that provide the necessary flexibility for executing image and speech processing algorithms.

REFERENCES

[1] Lewis, L., Amitai, Z., and Silverman, H. F. (1983). The APS-II processor for speech recognition. *1983 IEEE Int'l. Conf. Acoustics, Speech, and Signal Processing*, 483–486.

[2] Hockney, R. W., and Jeshope, C. R. (1981). *Parallel Computers*. Adam Hilger Ltd., Bristol, England.
[3] Flynn, M. J. (1966). Very high-speed computing systems. *Proc. IEEE*, **54**, 1901–1909.
[4] Siegel, H. J., Siegel, L. J, Kemmerer, F. C., Mueller Jr., P. T., Smalley Jr., H. E., and Smith, S. D. (1981). PASM: A partitionable SIMD/MIMD system for image processing and pattern recognition. *IEEE Trans. Comp.*, **C-30**, 934–947.
[5] Kapur, R. N., Premkumar, U. V., and Lipovski, G. J. (1980). Organization of the TRAC processor-memory subsystem. *AFIPS 1980 Nat'l. Comp. Conf.*, 623–629.
[6] Sejnowski, M. C., Upchurch, E. T., Kapur, R. N., Charlu, D. P. S., and Lipovski, G. J. (1980). An overview of the Texas reconfigurable array computer. *AFIPS 1980 Nat'l. Comp. Conf.*, 631–641.
[7] Siegel, H. J. (1984). *Interconnection Networks for Large-Scale Parallel Processing: Theory and Case Studies*. Lexington Books, Lexington, Massachusetts.
[8] Kuehn, J. T., Siegel, H. J., and Hallenbeck, P. D. (1982). Design and simulation of an MC68000-based multimicroprocessor system. *1982 Int'l. Conf. Parallel Processing*, 353–362.
[9] Siegel, H. J. (1983). The PASM system and parallel image processing. In *Computer Architectures for Spatially Distributed Data* (H. Freeman and G. G. Pieroni, eds.), Springer-Verlag, New York.
[10] Mueller Jr., P. T., Siegel, L. J., and Siegel, H. J. (1980). Parallel algorithms for the two-dimensional FFT. *Fifth Int'l. Conf. Pattern Recognition*, 497–502.
[11] Siegel, L. J. (1981). Image processing on a partitionable SIMD machine. In *Languages and Architectures for Image Processing* (M. Duff and S. Levialdi, eds.) Academic Press, London, pp. 293–300.
[12] Krygiel, A. J. (1976). An implementation of the Hadamard transform on the STARAN associative array processor. *1976 Int'l. Conf. Parallel Processing*, **34**.
[13] Siegel, L. J., Siegel, H. J., and Feather, A. E. (1982). Parallel processing approaches to image correlation. *IEEE Trans. Comp.*, **C-31**, 208–218.
[14] Warpenburg, M. R., and Siegel, L. J. (1982). SIMD image resampling. *IEEE Trans. Comp.*, **C-31**, 934–942.
[15] Stone, H. S. (1971). Parallel processing with the perfect shuffle. *IEEE Trans. Comp.*, **C-20**, 153–161.
[16] Siegel, L. J., Siegel, H. J., Safranek, R. J., and Yoder, M. A. (1980). SIMD algorithms to perform linear predictive coding for speech processing applications. *1980 Int'l. Conf. Parallel Processing*, 193–196.
[17] Yoder, M. A., and Siegel, L. J. (1982). Dynamic time warping algorithms for SIMD machines and VLSI processor arrays. *1982 Int'l. Conf. Acoustics, Speech, and Signal Processing*, 1274–1277.
[18] Mitchell, O. R., Reeves, A. P., and Fu, K-S. (1981). Shape and texture measurements for automated cartography. *1981 IEEE Comp. Soc. Conference on Pattern Recognition and Image Processing*, 367.
[19] Tuomenoksa, D. L., Adams III, G. B., Siegel, H. J., and Mitchell, O. R.

(1983). A parallel algorithm for contour extraction: advantages and architectural implications. *1983 IEEE Comp. Soc. Symp. Computer Vision and Pattern Recognition*, 336–344.

[20] Duda, R. O., and Hart, P. E. (1973). *Pattern Classification and Scene Analysis*. John Wiley and Sons, New York.

[21] Freeman, H. (1961). Techniques for the digital computer analysis of chain-encoded arbitrary plane curves. *Proc. NEC*, **17**, 421–432.

[22] Dijkstra, E. W. (1968). Cooperating sequential processes. In *Programming Languages* (F. Genuys, ed.). Academic Press, New York, pp. 43–112.

[23] Motorola Semiconductor Products Inc. (1979). *MC68000 16-bit Microprocessor User's Manual*, Motorola IC Division, Austin, Texas.

[24] McMillen, R. J., and Siegel, H. J. (1982). A comparison of cube type and data manipulator type networks. *Third Int'l. Conf. Distributed Computing Systems*, 614–621.

Chapter Ten

Real Applications on CLIP4

M. J. B. Duff

1. INTRODUCTION

The five workshops that have led to the publication of this book and to its three earlier companion volumes have been devoted to the three interdependent topics: languages, architectures, and algorithms for image processing. Contributors have emphasised the complex interrelationships among these factors that can lead to efficiency in processing, noting that efficiency can be measured in many ways and that systems judged efficient by one set of criteria may well be seen as grossly inefficient when other, equally sound criteria are evoked. A major difference in approach exists between the *computer theorists*, who rate efficiency in terms of full utilization of gates and memory, and the *computer pragmatists*, who are more concerned with processing speed/cost ratios. It is not the purpose of this chapter to take sides in the issue, nor to propose a new formal method for assessing efficiency, but rather to report the experiences of the Image Processing Group in the Department of Physics and Astronomy at University College London during the first three years of operation of the image processing laboratory based on CLIP4.

CLIP4 is a 96 × 96 single-bit processor array operating in a SIMD mode, usually on images derived from the central one-ninth of a standard, noninterlaced television frame (i.e., the central third of the middle 96 lines of each frame). It has been reported fully in the literature [1], [2] and will, therefore, not be described here. Instead, attention will be given to the various features of the hardware and software systems that support CLIP4 and make it possible to take advantage of its very fast processing capability. In later sections, some of the application programmes recently carried out will be described; and, finally, plans for the further development of the group's facilities will be outlined.

INTEGRATED TECHNOLOGY FOR
PARALLEL IMAGE PROCESSING
ISBN 0-12-444820-8

2. THE IMAGE PROCESSING HARDWARE

The hardware system supporting CLIP4 is shown schematically in Fig. 1. Although CLIP4 can operate autonomously, it is more convenient to use a conventional minicomputer as a host, because it allows easier communication between CLIP4 and its users, facilitates mass storage of digital images, and makes possible the use of high-level languages. Furthermore, the serial components of the image processing programs can be executed by the host, thus eliminating the need to over sophisticate the CLIP4 dedicated controller.

Images are supplied to CLIP4 from one of several *workstations*:

(a) *Illuminated table.* This employs both reflected and transmitted light illumination. The television camera can be raised or lowered and is equipped with a lens designed to give a range of magnifications of about 250:1, so that one pixel in the digitized image can represent anything between 25 μm to 6.25 mm in the object on the table (usually a photographic print or negative).

(b) *Microscope.* A conventional microscope with a micrometer driver X–Y stage, a rotating turret objective head, and a television camera mounted above the microscope, in addition to the binocular eyepieces, provides a means for direct digitization of images of microscope sections. A Dove prism

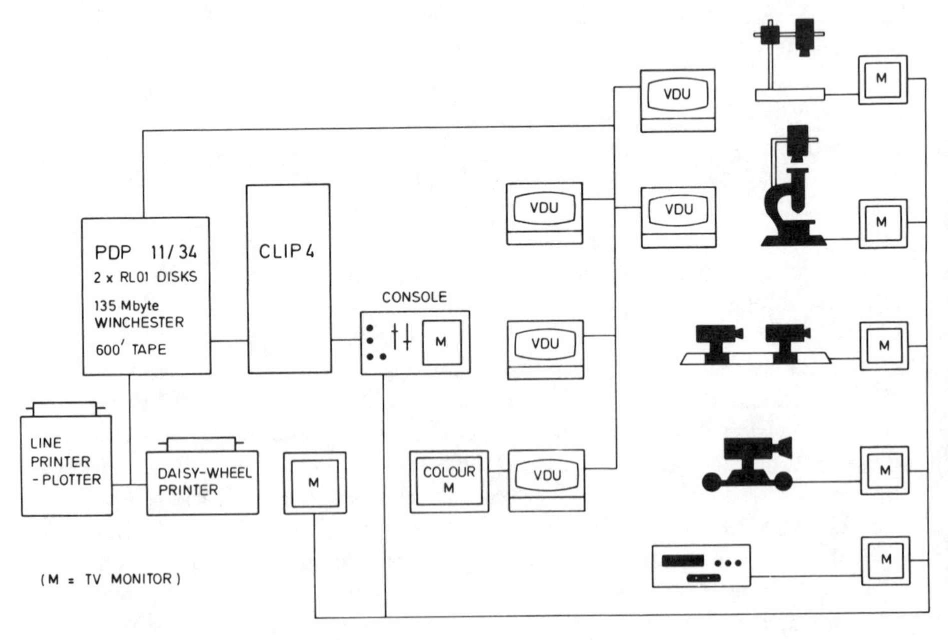

Fig. 1 The CLIP4 Laboratory Hardware System

is inserted in the optical path between the microscope and the camera so that images can be manually rotated prior to digitization.

(c) *Optical bench.* Two television cameras are mounted on an optical bench that is equipped with the usual range of attachments (lens holders, optical slits, etc.).

(d) *Camera trolley.* One camera is mounted on a trolley and fitted with a motorised pan and tilt head and a power zoom and focus lens. This unit is intended for use when large solid objects, rather than photographs, microscope slides, or other small articles are to be imaged.

(e) *Video recorder.* A Sony U-MATIC video cassette recorder is included in the system and can record or supply analogue images as required.

All the workstations are linked by means of a video distribution system. At each station, the user can switch the camera output to a local monitor and is then independent of the rest of the system. Alternatively, the processed image output from the CLIP4 console can be displayed. Switches at the console select which of the signals from the workstations are to be transmitted to CLIP4 for processing, and sliders control the relative brightness of the CLIP4 processed output and the original video signal to be mixed and distributed back to the workstations. Thus, it is possible for users to carry out preliminary visual inspection of images before occupying the processor.

Communication between the workstations and the PDP11/34 host is by means of Digital Equipment Corporation VT100 visual display units, some of which are permanently sited at the workstations. Unfortunately, the various image display modes are most easily controlled at the console (although most functions can be duplicated in software), so that some clustering of the workstations near CLIP4 is unavoidable. Further sophistication of the console could permit control from the VDUs but this is not yet a possibility.

Mass storage of images is provided at the PDP 11/34 on two 5-MB RL01 disks and a 135-MB Winchester disk. A small tape drive (600′, 800 bpi) can also be used as a backup or for transporting image data to or from the laboratories.

The production of image hard-copy output is always a problem. Four methods have been used so far; character overprinting using the Qume daisy wheel printer, dot-density methods based on the Versatec line printer–plotter, microfilm production using a microfilm plotter service in a central university computer facility, and direct photography of monitor screens using a Polaroid film camera. None of these methods has proved entirely satisfactory, although it is probably fair to say that the inherent low resolution of the 96 × 96 pixel data is more to blame for the disappointing results than are the hard-copy techniques. These will be tested more severely when 512 × 512 pixel data is introduced into the laboratory in the near future.

3. THE IMAGE PROCESSING SOFTWARE

An important aspect of the facilities offered in an image processing laboratory is the range of software presented to the user. The CLIP programme was hardware-driven during the early stages when the main objective was to devise a cellular array that would efficiently implement image processing operations. It could be argued that this approach would be more accurately described as algorithm-driven or operator-driven; but whichever term is used, it must be said that little attention was given to the design of image processing software.

More recently, however, spurred on by the completion of hardware that could be used to obtain real-time interactive image processing, considerable efforts have been made to improve the software environment. The following paragraphs outline the major features of the available software:

(a) *UNIX operating system.* CLIP4 is hosted by a conventional serial computer (PDP 11/34) that runs the UNIX version 7 operating system. The usual range of UNIX facilities for a serial computer are available to the user (file handlers, editors, compilers, text formatters, etc.) and will not be described here.

(b) *CAP code.* The assembly level language for CLIP4 is CAP, a mnemonic code assembling to 16- and 32-bit machine code instructions, closely resembling assemblers for conventional serial computers. The only unfamiliar instructions are the SET instructions which, when broadcast to the CLIP4 array, define the boolean operations to be performed by each of the dual boolean processors in every processing element in the array, and also define the interprocessor connectivity structure. An extensive CAP subroutine library has been written and is being enlarged frequently. These subroutines include efficient code for a comprehensive range of basic image processing algorithms (image arithmetic, filters, convolutions, histogram based operations, etc.), so that many complete programs involve little more than sequences of subroutine calls to the library.

(c) *Image-processing C (IPC).* The high-level language C, used extensively in the UNIX operating system, has been extended to deal effectively with images [3]. Image processing operations in the subroutine library described above are available as functions in C. Memory management has been incorporated; this is particularly important with CLIP4, because there are only 32 bits of storage associated with each processing element in the array. Further storage involves off-loading onto a Winchester disk. The memory management system ensures that space in the array is always made available for routines requiring fast access to certain image data, less immediately required data being automatically swapped out onto disk. The IPC subroutine library is more comprehensive than the CAP library, taking advantage of the serial processing capacity of the host as well as using CLIP4's parallel processors. None of the subroutines are completely optimal (in terms of execution speed), because efficiency is lost in providing generality of applicability and simplicity of use. For example, the user is not required to specify the length (number of bits/pixel) of the output image, and all the subroutines will accept input images of any length (up to a certain maximum). The majority of the programs designed by the Image Processing Group are written in IPC.

(d) *The Command system.* This system offers a small package of commands for image handling and inspection. Images can be written to or read from files, displayed numerically, and inspected using a keyboard driven cursor. Image segments can be written out numerically using the Versatec printer. Information can be listed describing all the images currently active in the program and indicating the availability of unassigned storage in the array. The operating system is arranged so that the command system can be *chained into*, i.e., a running program can be halted, the command system entered and used to inspect the various images in use in the program. It therefore provides a valuable program development tool.

(e) *MENU.* The MENU program is designed for the naïve user and for rapid assessment of newly acquired image data. The majority of the functions available in the subroutine library are executed as a result of performing two key-strokes on the VDU keyboard. For example, a new television image is read into the system and displayed on the console when *ip* is typed; the instruction *mf* operates on the displayed image and performs a 3×3 median filter, immediately displaying the result. In every case, the image displayed forms the input image, and the output image is then computed and appears in the display. Images can be named and stored temporarily in the array (if room is available), or on disk, or permanently in disk files. An interesting feature of MENU is that operations that are meaningful for both binary or grey images are called by the same instruction. Thus *eg* finds binary edges in binary images or grey edges (in which each pixel has a value proportional to the edge gradient) in grey images. The command v counts 1-elements in binary images or integrates pixel values (finds the grey volume) in grey images. The use of MENU can reduce the time needed for simple program development to a few minutes.

(f) *User programs.* All users of the system are encouraged to tidy up their programs and to make them generally available, either as subroutines or as complete programs. The catalogue of available software is, therefore, steadily expanding.

4. APPLICATION STUDIES

The array processor principle is not limited to full processor coverage of the image. Although it is convenient and efficient to provide one processor for every pixel in the image, an alternative and cheaper approach is to scan a small array of processors through the larger image area. This technique was applied successfully when CLIP3 [4], a 16 × 12 element array, was used to process 96 × 96 pixel images. The scanning process can be achieved without modifying the array structure, but it is worthwhile incorporating edge stores to assist in the joining together of the adjacent processed sectors.

The input/output hardware does need extensive adaptation if large images are to be scanned, and in fact, it may be found that a completely new system needs to be constructed. All the CLIP4 television cameras, the A–D and D–A converters, and the input and output image stores are set up for 96 × 96 pixel images. It was not considered worthwhile to attempt to rebuild the existing system to handle larger images; later CLIP systems are being designed *ab initio* with this in mind.

Much of the work that had been carried out using CLIP4 and the earlier CLIP systems had been directed towards obtaining an understanding of the structure of parallel algorithms, so the available image area was of little importance. The completion of the CLIP4 system offered the chance to undertake some application projects, but it was necessary to choose projects in which a 96 × 96 pixel image would represent a large enough image area at the required resolution. At the same time, it was recognised that higher resolution (larger area) systems would soon become available; thus, algorithms might be developed on CLIP4 and demonstrated to show feasibility, but with the understanding that production processing would be performed later when the new system had been commissioned.

Three projects were chosen as meeting the requirements implied by the current system: in none of the cases was there an immediate need for a commercially available CLIP system; all algorithm development could be sensibly performed within a 96 × 96 pixel image; the work could be performed in the Image Processing Group laboratory; an interested collaborator was available to provide expertise relating to the application rather than to the image analysis; each project was one in which the fast processing capability of CLIP4 would be of great value in the solution of the application problem. The three projects are outlined in the following paragraphs:

(a) *Tomographic reconstruction.* This project (a collaboration involving K. A. Clarke of the Image Processing Group and Dr. B. Goddard of the Nuclear Medicine Department at Southampton General Hospital) concerned the use of CLIP4 to reconstruct tomographic data from a radioactively labelled subject. Suitable substances such as sugars can be tagged radioactively and injected into a subject's bloodstream.

An Anger camera, placed beside the subject, forms a 2-D image of the radioactive material in the subject's veins (usually in the head or an organ such as the liver). The camera is then rotated around the subject into, typically, 72 different angular positions; so the complete data set is 72 photographs of the patient, viewed from each of the 72 angular positions.

If a horizontal strip is abstracted from each of these photographs, then the data in all of these strips are sufficient to reconstruct an image of the corresponding horizontal cross section of the subject, although, theoretically, an infinite number of views must be used to obtain a completely accurate reconstruction. Various techniques for reconstruction have been developed; the most common involves *back projection*, which has been implemented in CLIP4. The process can be visualized as setting up 72 projectors in the appropriate angular positions and using parallel optics to project through the strip photographs across the central region originally occupied by the subject. The intensities from all the projectors are summed throughout the reprojection space, and it can easily be deduced that a cross section image will form in the central region. A more careful examination of the process reveals the presence of reconstruction artifacts, which can be minimized by applying suitable convolution functions to the strip images before reprojection. The careful tuning of these functions is the major factor in determining the resolution available in the reconstructed images. Since each 2-D image comprises a large number of horizontal strips, a large stack of horizontal cross sections can be computed, constituting a complete 3-D image.

Parallelism in CLIP4 is exploited in several ways in this process. A set of strips, selected from each image, can be stacked together to form a new image on which a single convolution can be applied in one go. Reprojection through a strip can be carried out simultaneously for each element in the strip, using the global propagation property of the array in which the image data are made to move along paths at angles determined by the reprojection position. Summation into an output image array is performed in parallel at every element of the array. The complete process is fully described by Clarke [5]. Figure 2 shows stages in the reconstruction of some data produced artificially for assessment of reconstruction resolution; the low definition in these images is intrinsic in emission tomographic data.

The data obtainable by emission tomography are inherently not well resolved, so 64 × 64 pixel images were considered appropriate. The reconstruction time (for a single cross section) using CLIP4 and extrapolating to an optimised program using CLIP4 equipped with 4B circuits (recently

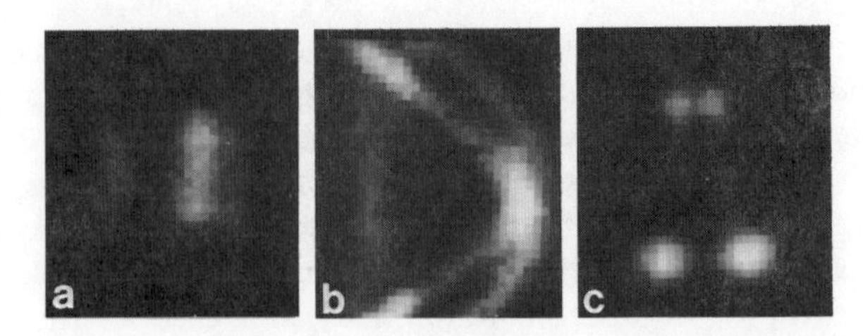

Fig. 2(a) A side view of the subject (a phantom consisting of six vertical tubes labelled radioactively) at a particular viewing angle. (b) A composite image formed by putting together rows of pixels taken one each from the images at every viewing angle. (c) The reconstructed cross section formed by back-projecting the image rows in Fig. 2 (b) after application of a suitable convolution function.

available integrated circuits meeting the original full specification; 4A circuits had to be run at 0.4 times their intended speed, due to manufacturing problems) is approximately 150 ms, a figure comparing favourably with hard-wired special-purpose computing circuits for reconstruction.

(b) *Electrophoresis gel analysis.* This project involved D. J. Potter of the Image Processing Group and various staff of the Imperial Cancer Research Fund Laboratories, in particular Dr. L. Franks. Proteins can be made to spread out across thin gelatine sheets in two directions mutually at right angles and in the plane of the sheets. The displacement in one direction is proportional to the molecular weight of the protein molecule and, in the other, proportional to the molecule's isoelectric point. The displacements are caused by a complex diffusion process in a variable *p*H medium under the influence of an applied electrostatic field. The positions of the proteins after diffusion can be made apparent by radioactive tagging or by staining. Ideally, each protein would be represented by a small, isolated dot, whose density would be proportional to the amount of protein present. In practice, every dot is blurred by being smeared in the two orthogonal directions, and adjacent spots often touch or even overlap.

Thus, the electrophoresis gel provides a form of map of the proteins in the sample being investigated. The gels are subject to distortion which, with the smearing effect, makes it impossible to relate spot coordinate position directly to protein type. In the CLIP4 project, therefore, an attempt was made to reverse the effects of smearing and gel distortion so that spot coordinates could be compared with those on a calibrated gel.

The smearing effect was assumed to be uniform in every region of the gel so that a deconvolution could be applied uniformly over the whole image, thus fully exploiting the array's parallelism. The gel distortion was removed by using known protein calibration points to compute a low-order distortion transformation, matching residual unmatched points by applying higher-order corrections to bring nearby groups into correspondence. Stages in this process are illustrated in Fig. 3, but the complete procedure has not yet been reported. This process is now being used in the routine examination of electrophoresis gels.

(c) *Carotid body reconstruction.* H. H.-S. Ip of the Image Processing Group has been collaborating with Dr. J. A. Clarke and Professor M. de Burgh Daly of St. Bartholomew's Hospital Medical College in a project to

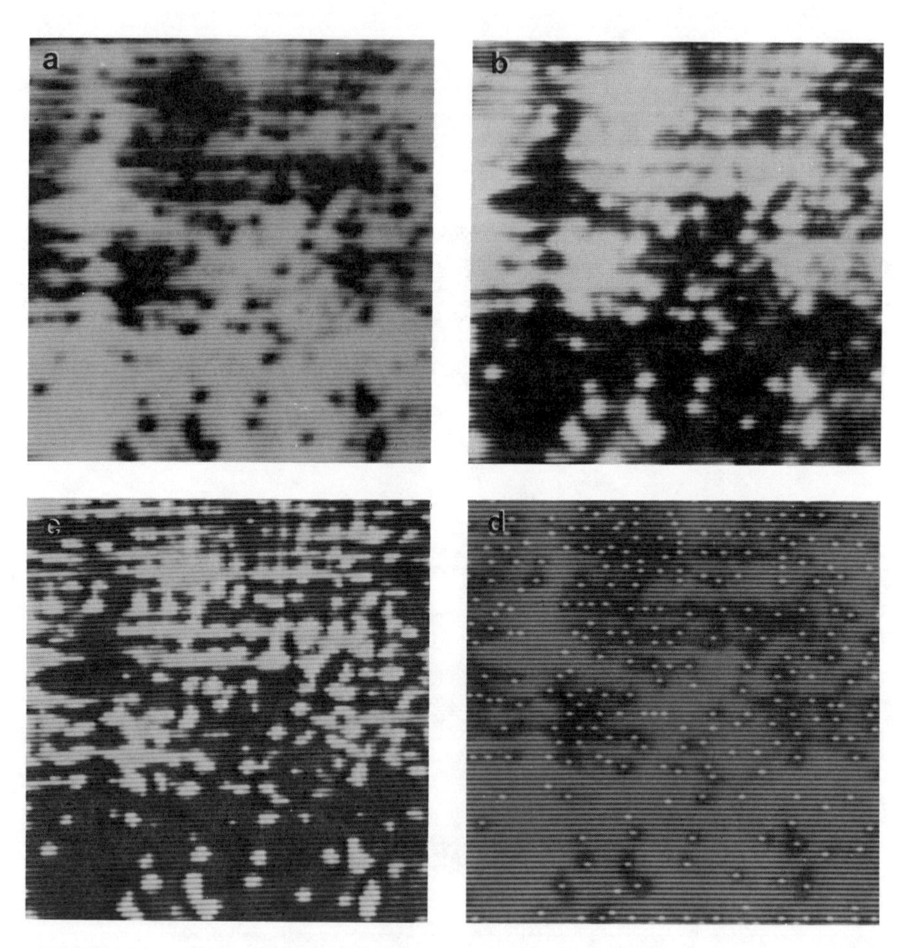

Fig. 3(a) A portion of an electrophoresis gel. (b) The gel after high-pass filtering. (c) The same gel after enhancement by convolution with a specially tuned convolution kernal. (d) Identification of spot centres for comparison with a calibrated prototype.

investigate the vasculature of the carotid body. This small organ (a pear-shaped object about 5 mm long in the rat) is sliced into some 250 2 μm sections that are then viewed under a medium-power optical microscope. The cross sections of the many small blood vessels forming the vasculature are clearly visible, but it is difficult to form an impression of the structure of this complex network of vessels by examining consecutive slices.

In the CLIP4 project, the first task is to align correctly the slices one-by-one under the microscope, so that the images can be digitized in a

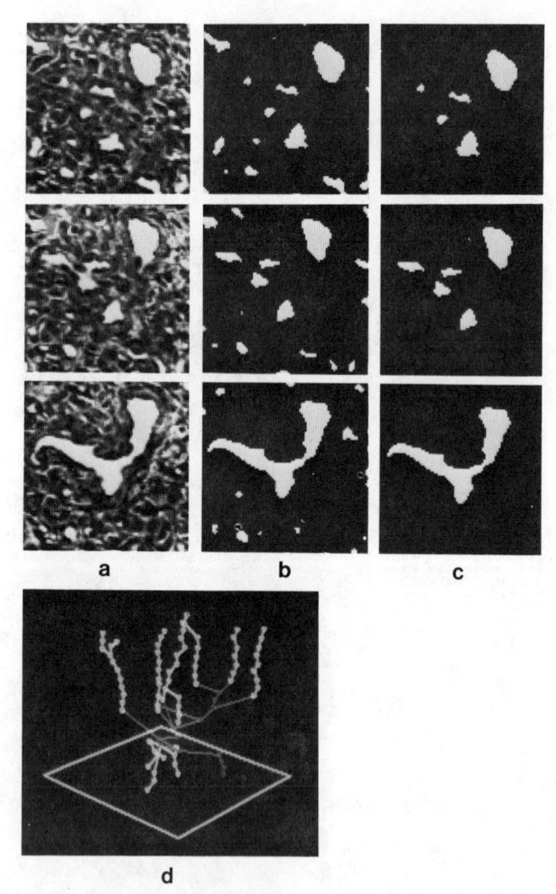

Fig. 4(a) Sections of the carotid body. (b) The same sections showing the principal blood vessels. (c) Vessels which connect to others in adjacent sections. (d) Displays showing connected vessels forming a vasculature.

correct relationship. This is achieved by extracting the edges of the vessels, finding the centres of gravity of the edge images, and shifting the second image so that the two centres coincide. The second image is then translated and rotated so as to minimize the differences between the two edge images (using the exclusive OR function). It is then a comparatively simple process to move the second microscope slide so that it aligns with its computed shifted image. It is hoped to improve this operation by employing a motorized stage at a later date. Note that the manual shifting followed by redigitization eliminates the small amount of blurring caused by the computer shifting operation.

Various feature-dependent tests are used to eliminate spurious edges, and then a stack of edge images is processed to trace through connected structures, taking further opportunities to eliminate edges that do not form part of connected structures. The system then outputs a graph-structured list that can be used to construct one of several alternative perspective views of the vasculature or of systematic representations of it. The complete process has been reported in detail in [6].

One of the major problems in this project has been the need to operate on a very large number of images, necessitating extensive use of disk storage and multiple swappings in and out of the array. Apart from this, all the processing leaned heavily on the utilization of the parallel capability of CLIP4. The method is obviously extendible to other organs and tissue structures.

5. COMMENTS ON THE APPLICATION STUDIES

Each of these projects has served to evaluate the CLIP4 system under realistic operating conditions, highlighting its weaknesses and its strengths. On the positive side, the parallel structure of CLIP4 makes satisfactory operating speeds possible, both during the usually highly interactive algorithm and program development stage and, eventually, during production processing. The workstation concept has been convenient as implemented, and the sub-routine libraries and common programming language (IPC) have stimulated program sharing and helpful discussions.

On the negative side, the very limited in-array storage (32 bits/pixel) has repeatedly degraded performance or inhibited the choice of algorithm, and the small image area (96×96 pixels) has meant that in two of the projects (electrophoresis gel analysis and carotid body reconstruction) the techniques developed cannot yet reach their full potential. Nevertheless, algorithm development was not affected by the image size limitation.

The carotid body problem showed the weakness in our graphics capabilities. Although all the information necessary to reconstruct 3-D images of the relevant structures had been extracted, the required display hardware and software were not available.

Finally, it became clear that better quality television cameras could yield substantially superior results but that resolution would then be limited by inadequate microscope optics, once 512 × 512 pixel images were to be produced.

One of the above factors, which cannot be overemphasised, is the advantage gained by working with an image processor that gives an apparently instantaneous response during algorithm development. Such a rapid response is very conducive to productive thinking, so much so that for those who have experienced it, a reversion to developing programs on a batch-processing computer would seem quite intolerable. It is also worth commenting that execution times of a few tens of seconds can be exceedingly frustrating, more so than delays of an hour, which at least permit alternative activities. This factor is the main argument against the use of some of the cheaper but slower image processing systems.

6. FUTURE DEVELOPMENTS

The next two years will be used to improve the Image Processing Group's facilities while continuing with a range of application studies. Three new machines are planned and are in various stages of development. At the time of writing (October 1984), CLIP4S is being debugged and should be in full operation shortly. It is built around an array of CLIP4B processors that scan and process 512 × 512 pixel images. CLIP4S will run CLIP4 programs substantially unchanged and is, therefore, a slow, high-resolution backup for CLIP4. It is intended that program development should still be performed on CLIP4, CLIP4S being used only for production processing.

At about the same time, CLIP4R should be commissioned. This is a version of CLIP4 constructed by the Rutherford Appleton Laboratory, incorporating some 170 bits/pixel of 40-μs access memory and intended to interface to a 512 × 512 framestore. The 96 × 96 array of processors will extract data from program selectable locations in the framestore. Programs, apart from input/output, should be CLIP4 compatible.

The most advanced processor being constructed is CLIP7, based on a custom-designed integrated circuit including a 16-bit ALU. This is described by Fountain in this volume (Chapter 13) and will not be discussed here. However, it is of interest to note that, among other things, the CLIP7 is expected to provide processing speeds for 512 × 512 images approximately equivalent to speeds on 96 × 96 images achieved in CLIP4.

All these systems are to be linked through a local-area network, and a high-resolution (512 × 512) video distribution system will also be provided. Funds are being actively sought to upgrade the workstations and to acquire a high quality colour-graphics facility. If all these plans materialise, the weaknesses found during the current application studies should be largely eliminated.

REFERENCES

[1] Duff, M. J. B. (1978). Review of the CLIP image processing system. *Proc. 1978 Nat. Comp. Conf.* 1055–1060.
[2] Fountain, T. J. (1981). CLIP4: A progress report. In *Languages and Architectures for Image Processing* (M. J. B. Duff and S. Levialdi, eds.). Academic Press, London, pp. 283–291.
[3] Reynolds, D. E., and Otto, G. P. (1982). IPC User Manual. Image Processing Group Report No. 82/4, *Department of Physics and Astronomy*, University College, London.
[4] Duff, M. J. B., and Watson, D. M. (1975). CLIP3: A cellular logic image processor. In *New Concepts and Technologies in Parallel Information Processing* (E. R. Caianiello, ed.). Noordhoff Int. Pub., Leyden, pp. 75–86.
[5] Clarke, K. A. (1984). *Reconstruction of Nuclear Medicine Images on the CLIP4 Computer*. Ph D. Thesis, University of London.
[6] Ip, H. H.-S. (1983). Detection and three-dimensional reconstruction of a vascular network from serial sections. *Patt. Recog. Lttrs.* **1**, Nos. 5, 6, 497–505.

Chapter Eleven

The Application of Three-dimensional Microelectronics to Image Analysis

Graham R. Nudd, Jan Grinberg, R. D. Etchells, and M. Little

1. INTRODUCTION

Over the past several years, researchers in the fields of image analysis and understanding have become convinced that the development of fully automatic vision systems for applications such as object identification and pattern recognition are realistic goals. Certainly the benefits of such systems are well recognized in the military arena (e.g., for target acquisition and autonomous guidance) and in the commercial world (e.g., for product inspection and robotics control). A primary difficulty in the development of such systems has, until very recently, been the large throughput required. We now anticipate that the developments currently under way in Very Large Scale Integration (VLSI) of microelectronics will provide a solution to these problems. Certainly as microelectronic devices decrease in size below the one micron level, for example, the inherent gate delays and associated switching speeds become more favorable. However, it is extremely unlikely that these effects alone will enable the necessary computing power to be achieved without radically different architectures. Accordingly, we are now beginning to see a great deal of activity focused on improving existing architectures for image analysis. Further, the importance of this application has initiated a considerable effort to develop new VLSI architectures that are optimized for image based problems. For example, considerable effort has been expended in developing cellular machines such as the Cellular Logic Image Processor (CLIP)[1] and the Massively Parallel Processor (MPP)[2]. These are well

INTEGRATED TECHNOLOGY FOR
PARALLEL IMAGE PROCESSING
ISBN 0-12-444820-8

matched to the 2-D nature of imagery and are essentially bit-serial arrays, ideally with one processing element assigned to each pixel. Other work has been concerned with the development of 2-D systolic arrays for specialized preprocessing operations [3]. However, the problems associated with memory access, data loading, and intercommunication of the processors present significant limits on these types of architectures. Indeed, it has now been recognized that the access and communication issues are the principal barriers to the successful exploitation of VLSI concurrency.

We discuss here the potential application of 3-D microelectronics (3-DM) for solutions to these problems. After a brief review of the underlying microelectronic technologies required to make 3-DM possible, we summarize the potential impact from an architectural viewpoint and give an example of an implementation of a Single Instruction Multiple Data (SIMD) architecture that we are now working on for both image analysis and other applications.

2. TECHNOLOGY ISSUES

Some of the most significant problems in the design and layout of complex VLSI chips are associated with the interconnection and routing issues. Typically, in high-density processors, the silicon area devoted to device interconnects far exceeds that occupied by the active devices. Also, the two key parameters of effective switching rates and on-chip power consumption are dominated almost entirely by the efficiency of the routing. One would, therefore, anticipate that considerable benefits might be derived if microelectronic circuitry could be built in a three-dimensional medium. For example, devices might be clustered closer together, and the length of routing interconnects, buses, and so forth might be shortened simply by moving over or under intervening cells. The actual savings, in terms of wiring length and silicon area, depends on the nature of the specific circuits being implemented, but upper and lower bounds for a number of widely used circuits, such as permutation networks, etc., can be calculated. In a recent paper, Rosenberg [4] provides estimates of the saving in both wiring and silicon achievable by 3-D mapping. In fact, it can be shown for the simplest case of random interconnection schemes (a case one would wish to avoid, of course) that the average length of interconnect per node increases as N_T (the total number of transistors per chip) for the 2-D case, but only as $N_T^{1/2}$ for the 3-D case. This saving, of course, can partially relieve the dilemma that as devices get smaller, increasing the circuit density and therefore N_T, they are required to deliver more energy to drive the proportionally increased interconnect length. Significantly greater benefits can accrue when the memory access issues are taken into account, as in our cellular architecture discussed below.

Until recently there has been no effective way of exploiting the proposed benefits of 3-D–configured architectures. Work was initiated by Duff using printed circuit boards and a square array of pin interconnects, one for each pixel. However, with this approach, the length of the communication path between adjacent boards would produce a highly anisotropic array with today's technology and would prevent the above projected benefits from being obtained. More recently our work at Hughes Research Laboratories has concentrated on a stacked silicon wafer concept that uses proprietary interconnect and wafer feedthrough techniques. The resulting 3-D computer is hybrid package, as illustrated in Fig. 1; but the spacing between adjacent wafers is 20 mil or so and hence is comparable to the sizing of the 20 mil × 20 mil processing elements on each wafer. Two underlying technologies are required to achieve this configuration: a highly conductive means of passing signals through each wafer and a means of interconnecting adjacent wafers. We use a single feedthrough and interconnect at each PE site, which results in an $n \times n$ array of feedthrough–interconnect combinations. Each interconnect consists of a highly doped p-type via (one for each PE) running through the wafer. A matching array of micro interconnects, one at each feedthrough site, provides electrical conductivity through the entire wafer stack, as shown in Fig. 1. In this way we are able to form an $n \times n$ array of microprocessors distributed in the z direction, one behind each pixel. Our initial machine is designed around a PE cell size of 20 mil × 20 mil and a 32 × 32 array, but we are planning arrays of up to 512 × 512 elements to match high quality imagery.

The feedthroughs and interconnection capabilities can be formed by a variety of techniques including laser drilling and indium bump technologies.

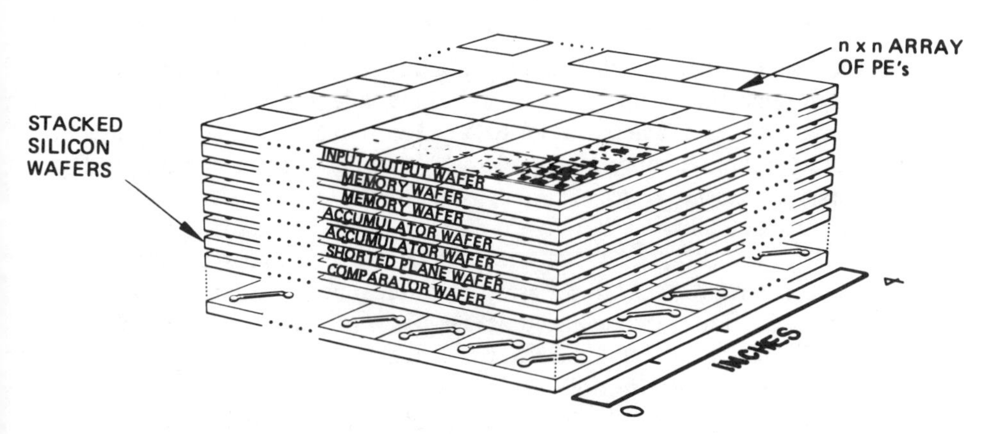

Fig. 1 Concept of HRL 3-D computer for image analysis

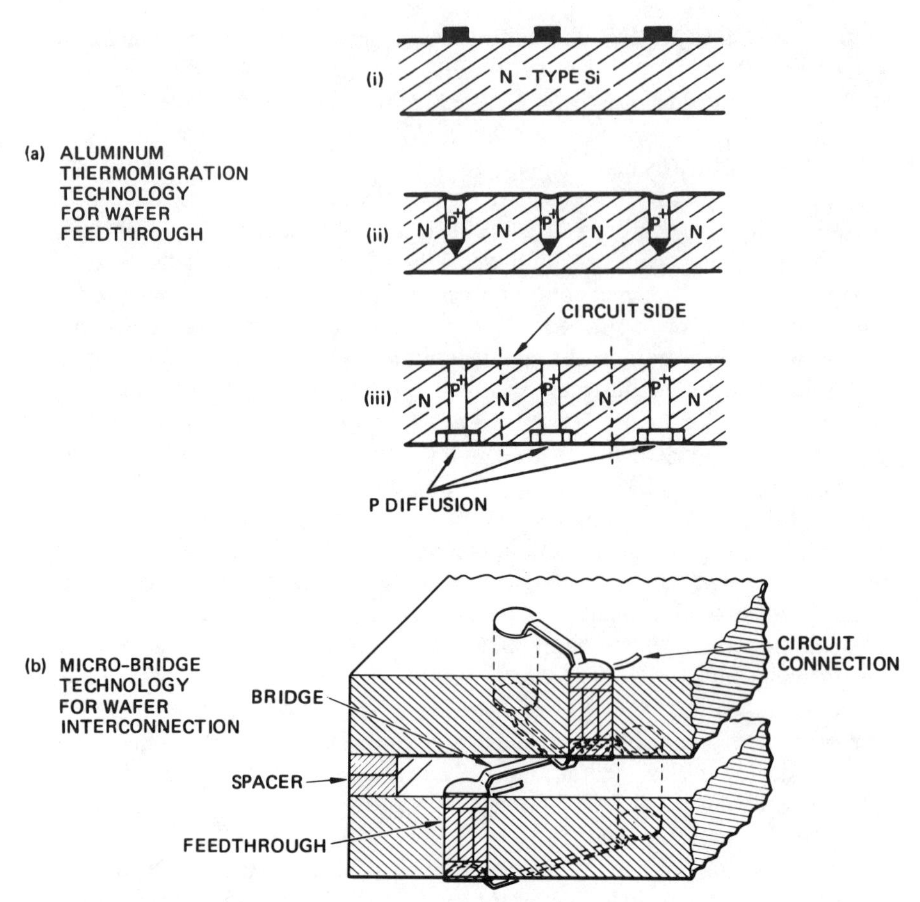

Fig. 2 Basic technology for the 3-D computer

However, we use thermal migration of aluminum to provide a heavily doped p^+ trail through the silicon. In this way we are able to provide buses with approximately 1-mil diameter for all 1024 cell sites in a single planar process compatible with other conventional microelectronic processing. A number of proprietary processing issues [5] are involved in our current technologies so that we are able to obtain a 100% yield over the full 32×32 array, resulting in the necessary 1024 serial bus lines. We also use a planar evaporation process to produce the 32×32 array of interconnects required between adjacent wafers. These interconnects are necessary to provide for mechanical contact, taking into account distortions on the wafer, and good electrical contact. At present the flexible bridge structure we employ can provide deflections of the order of 10 μm to assure satisfying both the above requirements. Details of these processes have been published in previous publications [6], [7] but are summarized in Fig. 2. A resulting wafer with the feedthrough, interconnects, and an array of CMOS processed circuits is shown in Fig. 3. The development of the full physical structure for the 3-DM architecture from this point thus becomes an issue of assembling sufficient processed wafers to form the stack.

There are a number of significant advantages resulting from the hybrid approach described here, in which each layer is on a separate wafer. First, it is significantly easier to manufacture than any of the proposed monolithic structures, in that relatively high yields can be obtained by testing at each stage throughout the manufacture. This includes the feedthroughs, interconnects, and the full circuitry on the individual wafers prior to the final assembly. Further, the processing of a subsequent layer does not affect the previous levels as it might in a fully monolithic approach. However, it is possible that a fully monolithic approach along the lines of the silicon-on-

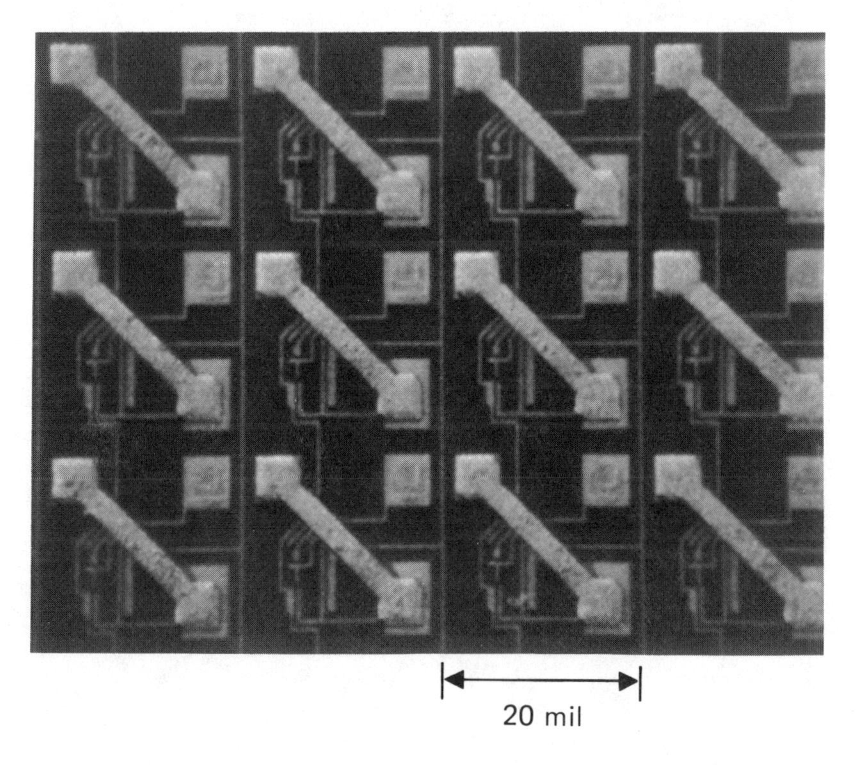

Fig. 3 Portion of assembled wafer with feedthrough, interconnect, and processed CMOS circuitry

insulator (SOI) concepts currently being developed [8] will become a viable technology for this architecture. One concept presently being pursued is aimed at growing amorphous or polycrystalline silicon over the final protective oxide (typically included on microelectronic circuits) and then recrystallizing this to form a high quality single crystal layer on which a successive layer of circuitry can be developed. This process would then continue on successive oxide layers until sufficient levels had been developed to form the full machine. Currently, laser beams, strip heaters; and electron-beam techniques are being investigated for the recrystallizing process, all with some success. The technology is presently in the early research stages and some years away from practical applications, but it might be a viable technology in time.

The essential issue is that, independent of the ultimate technology, the added dimension of 3-DM allows a number of significant opportunities and advantages over conventional microelectronics, and substantial architectural advances can be made independent of the ultimate implementation technology. Any significant differences will probably come at the device level, at which we may see new device types such as single gates distributed in the vertical direction. However, it is significantly above this level that the most interesting and significant architectural issues exist, and in this region the various technologies offer analogous opportunities.

3. ARCHITECTURAL ISSUES

As an illustration of the flexibility provided by 3-DM, we consider the machine we are currently developing for a wide variety of image analysis applications. The basic architecture can be classified as a cellular SIMD array. Our present array consists of 32 × 32 identical processing elements per wafer. Each wafer contains one particular type of processing element (e.g., an array of memory registers, accumulators, and comparators), ideally one for each pixel in the image window. A stack of varying wafer types is then assembled to provide an effective microcomputer behind each pixel and form the full machine. A consequence of this structure is that only as many wafers as are required for any particular application need be assembled, resulting in significant cost savings. A schematic view of the structure is shown in Fig. 4, this time looking at the stack edge-on. The control processor shown is used to exercise the array hardware and is shown as a separate noncustom unit. In this illustration, the silicon wafers are represented by the vertical rectangle. The horizontal lines running through the stack depict the data and control buses. The architecture is word-parallel and bit-serial in that while the logic of the individual processors in the array employs serial arithmetic, all of the

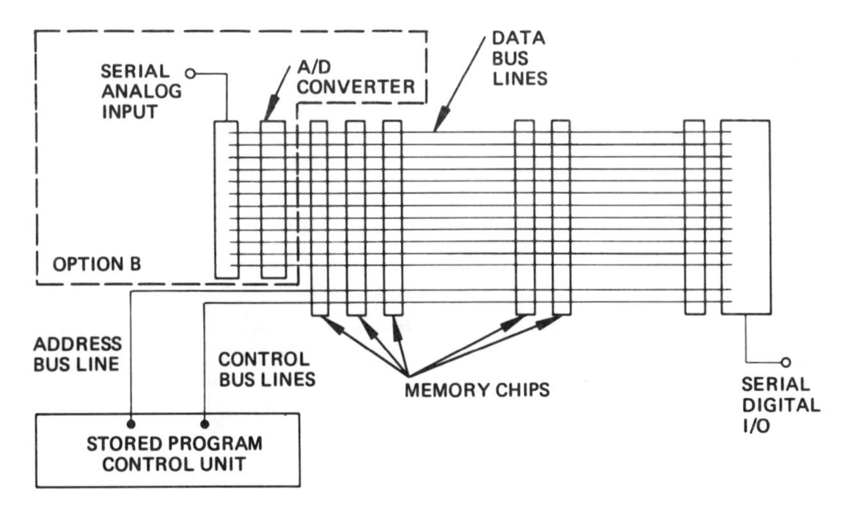

Fig. 4 Schematic of 3-D computer showing parallel bus structure from side view

processors operate simultaneously, in a word-parallel fashion. This large degree of parallelism more than compensates for the slower bit-serial arithmetic.

A principal advantage of the concept is that the logic gates forming each PE can in this way be made very simple. At present we are using only about 100 transistors per cell. This significantly reduces the problem of providing acceptable yield over the full wafer. For example, it can be shown that if we use a 2:1 redundancy at the cell level, the overall yield Y_T for a full $N \times N$ array of N_T transistors (assuming uncorrelated defect errors) is given by

$$Y_T = [1 - (1 - P^m)^2]^{N_T/m}$$

where P is the yield for a single transistor and m is the number of transistors per PE. Using present day yield figures and a 100- to 200-transistor block for each PE, one can calculate that acceptable wafer yield (20–30%) can be obtained using 2:1 redundancy at each PE.

Although the basic machine has an SIMD organization, implying that all processors in the array work in lockstep under the direction of the control unit shown, provision is included for data dependent branching operations. This is achieved by using a masking plane to select any desired subset of the image, and the circuitry for this is illustrated in Fig. 5. In normal operation a full plane of data (1024 pixels in our initial machine) is directly transmitted in parallel on the 1024 bus lines to any desired logic plane. By loading a desired pattern on the masking plane, any data subset can be chosen. It implements a wired-*and* function at each cell, as can be seen from the schematic, so that

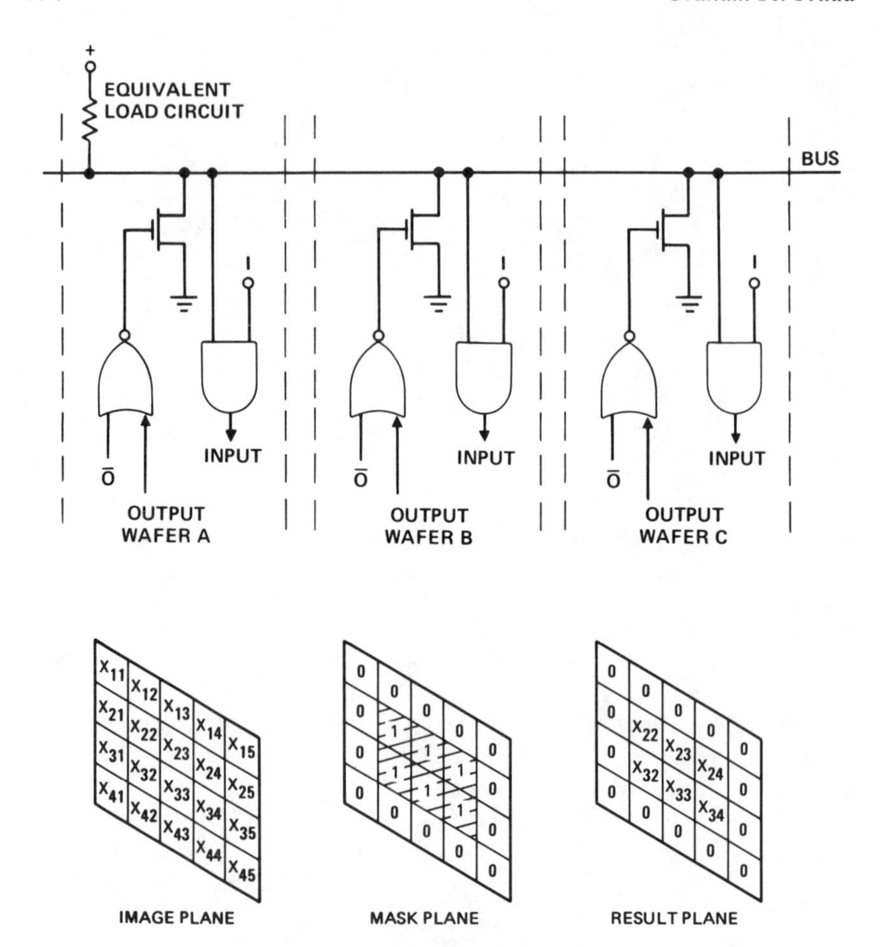

Fig. 5 Concept of masking operation using wired-*and* bus connect

data can only be transmitted along the bus if all cells on that bus either are in the *off* (no output) state or are transmitting a logical *one*. A zero presented by any of the circuits on a given bus line will pull that line low, forcing a zero to be input to any of the circuits accepting data. In image analysis, the binary control masks representing, for example, interior versus exterior points or local threshold operations, occur frequently as a natural part of the intermediate processing routines. In these applications the wired-*and* function is effective means of segregating object and background data, for example, for separate processing.

In the normal mode the processing is controlled by a stored program residing in the control unit, and computations are executed as planes of data move up and down the stack to any preselected wafer location. Each wafer contains an address decoder and instruction, or configuration latch connected to these lines. Prior to the execution of any instruction by the array, all wafers are in a neutral state, in which they do not communicate with the data buses and ignore the system clock signals. In preparation for an operation, the control processor configures each wafer involved by transmitting its address over the address bus and the appropriate configuration code over the control bus. When a wafer recognizes its address, it strobes the data currently on the control bus into the on-chip instruction latch. In this way all the wafers to be used in a particular operation are configured sequentially, prior to the start of the operation. Once the necessary wafers have been so configured, the control unit passes the appropriate number of clock pulses for each operation to the stack hardware. After the operation is performed, the control processor returns the stack elements to their neutral state by toggling a reset line common to all wafers.

The technology used to implement the circuitry is a combination of CMOS and CCD. The clock frequency is 10 MHz, and our simulations indicate that an average of 2.5 wafers are active at any one time. Consequently, the overhead associated with the sequential configuration of the individual wafers is roughly 18%. This figure corresponds to the use of an average of 2.5 clock cycles for wafer configuration and one cycle for stack reset. As presently configured the word-length is 16 bits, and hence 16 cycles are devoted to the actual computation itself. Thus, 3.5 out of 19.5 cycles are lost to control operations.

For most of the processing algorithms simulated to date, we have found a total of five wafer types to be sufficient, as indicated in Table 1.

Table 1
List of cell types in the 3-D computer

Cell type	*Function*
Memory	Store, shift, invert/noninvert, *or*, full word/MSB only, destructive/nondestructive readout
Accumulator	Store, add, full word/MSB only, destructive/nondestructive readout
Replicator plane	I/O, X/XY short, stack/control unit communication
Counter	Count in/shift out
Comparator	Store (reference, greater/equal/lower)
Multi *and* on bus line	

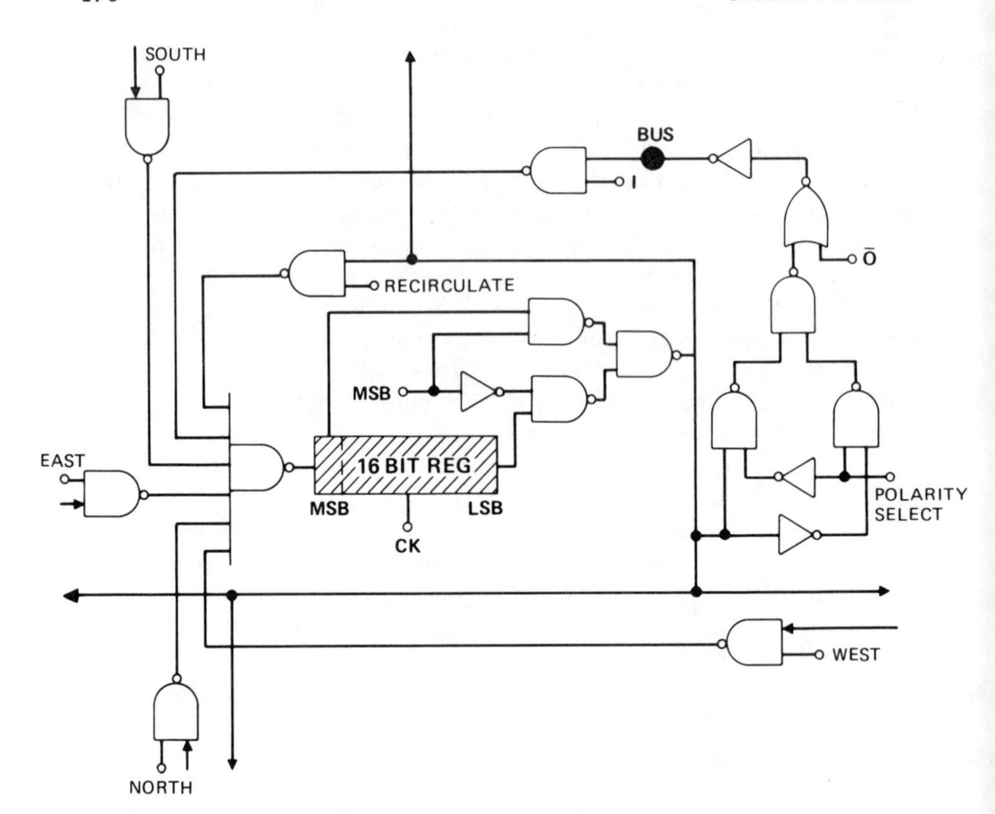

Fig. 6 Memory cell circuitry

The memory and accumulator cells are used to perform most of the basic operations in the machine. As shown in Fig. 6, each memory cell has the capability for communicating laterally in the plane to its north, east, south, and west neighbors (without destructive readout) and through the wafer to any of the cells along its bus line. Data may be logically inverted to facilitate two's-complement subtraction, and the option also exists to read out either the entire contents of the register or just the most significant bit. This capability is useful for sign testing and logical masking, as discussed above.

The accumulator circuitry is similar to that of the memory, differing in that there is no nearest-neighbor communication, and a one-bit serial adder is included. Two's-complement subtraction is provided by allowing a *carry* to be introduced into the least-significant bit of the word. A schematic of the accumulator circuit is given in Fig. 7. Multiplication is performed in the array by an adaptation of the classical shift-and-add approach so that a full 8-bit × 8-bit product can be formed in approximately 40 μs.

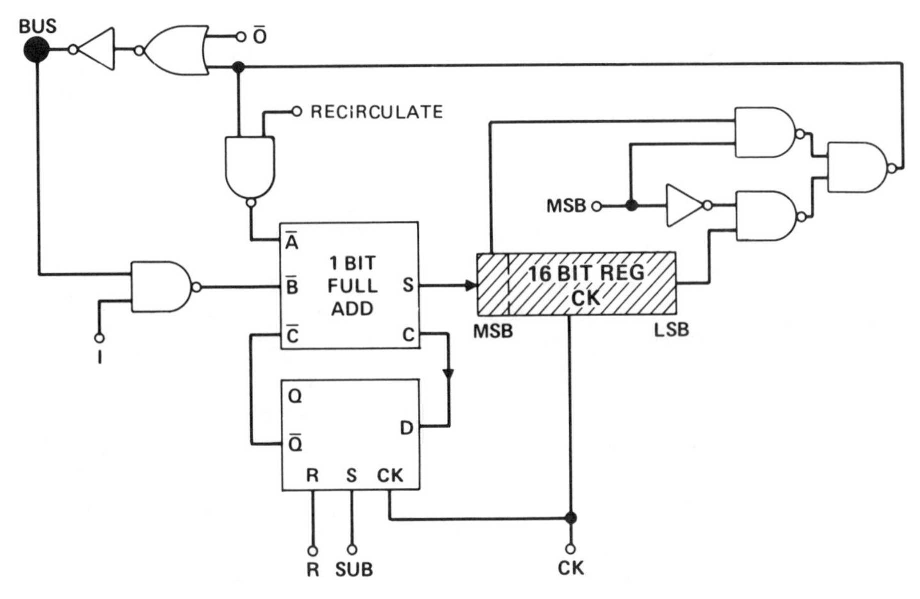

Fig. 7 Accumulator cell circuitry

It is advantageous, though not essential, to include a replicator plane in the architecture. Its function is to broadcast rapidly the constants and single values across the entire plane. It can be operated in two modes, by feeding data from the topmost row and propagating either in the vertical direction or row-wise. In either case, the overall effect is to optimize the speed at which single values can be communicated to all cells. Typical applications occur in thresholding and gain control modes.

The counter and comparator planes are useful (although again not essential) for image analysis. Typically, the counter is used in statistical operations such as event monitoring or histogramming. An internal 5-bit serial register is contained in each cell for data accumulation and subsequent readout. The comparator plane is used to perform serial comparisons of 16-bit words taken bit by bit from the bus with local stored 16-bit words. The three results—greater than, equal to, or less than—are stored locally for all cells, and three cycles are required to interrogate for the final result.

The circuit details of these logic cells can be found in Tsaur *et al.* [8]. Despite the bit-serial nature of the processing, the massive parallelism at the cell level provides for high execution rates. In Table 2 we provide the processing time required for arithmetic operations over the full $n \times n$ array.

Table 2
Processing times for primitive arithmetic operations

DATA MOVE	MEM	→MEM	1.8 μsec
ADD	ACC + MEM	→ACC	1.8 μsec
MULTIPLY	ACC × MEM	→ACC	42.2 μsec
DIVIDE	ACC ÷ MEM	→ACC	127.1 μsec
SQUARE ROOT	$\sqrt{ACC}$	→ACC	152.6 μsec

4. PROCESSING APPLICATIONS

For image analysis and other 2-D data processing, the maximum advantage can be obtained from the 3-D architecture when an individual processing cell can be assigned to each pixel. The processing thus commences by feeding a full frame or subwindow of the image onto a single wafer in a direct iconic format. In this way, on one memory plane of the stack we have a copy of the original image. The input/output interface to achieve this is essentially a series/parallel feed in which the input of line-scanned data is fed a line at a time to the uppermost row of a memory plane and then shifted in parallel, line by line, throughout the wafer. We are aiming at a capability for accepting data at 100 MHz rates for the line-scan portion and achieving parallel transfers of approximately 1.0 MHz. Once this operation has been completed, certainly for all the low-level feature extraction type operations, there is no need to reference bulk or secondary memory. Data access simply implies passage of the data in parallel through the stack to the appropriate processing wafer. We alluded to the set-up procedure and instruction flow for this earlier, and we now illustrate the processing concepts by schematics of the data flow rather than the detailed instruction by instruction program steps.

In the course of the development program for the 3-D computer, detailed simulations and code have been generated for a wide range of applications including matrix manipulations, synthetic array radar, feature extraction and classification. In the space available here, it is impossible to cover all these, so we attempt to illustrate the processing power by reference to three generic functions common to image analysis.

4.1 Two-dimensional spatial filtering

Local area filtering functions represent a very important class of operations in low-level image analysis and preprocessing of 2-D data. The simplest of these operations might be the calculation of local average intensities for the entire image for subsequent local area gain and brightness control or

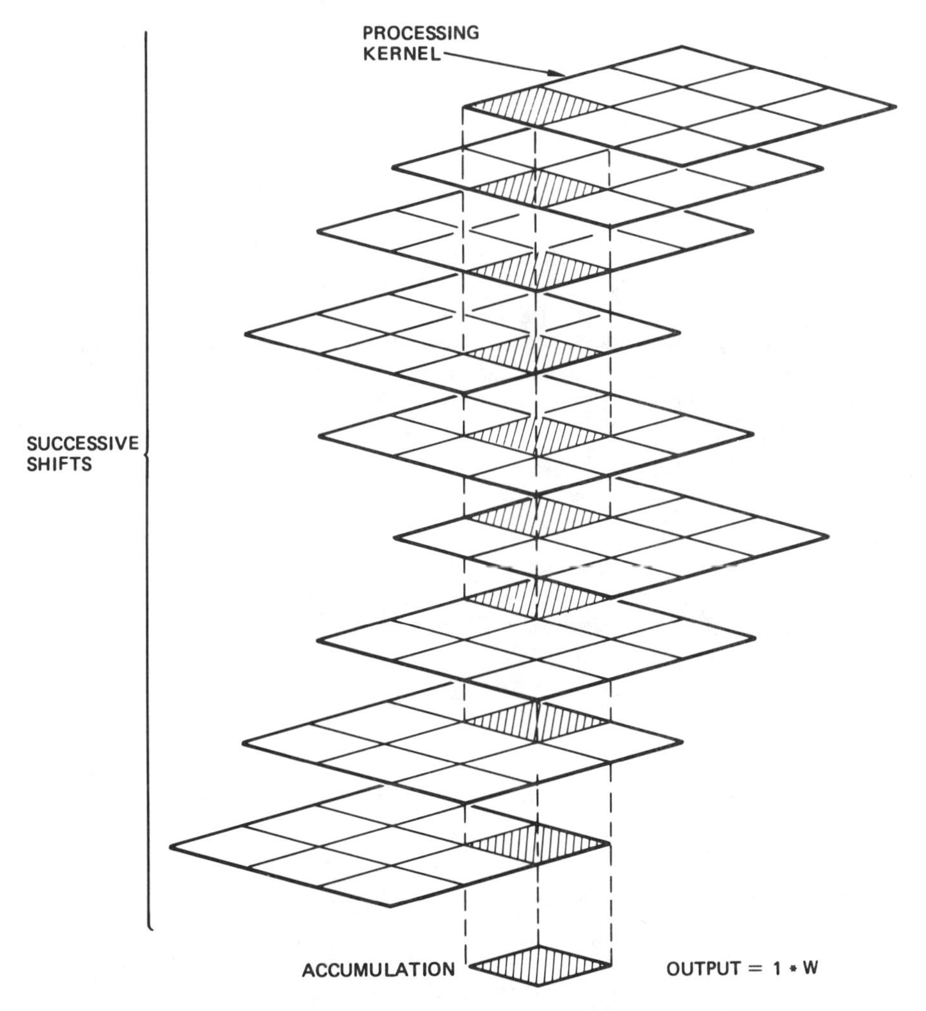

Fig. 8 Concept of local-area filtering operation by successive shifts and accumulation

thresholding. Alternatively, operations such as edge detection, high- and low-pass filtering, or gaussian weighting of some form are frequently used at the feature extraction level of image analysis. All these operations can be broken down into some form of sum of products or local convolution of the form

$$\tilde{I} = I * W$$

where I is the original image, $\tilde{I}$ is the processed image, and W is a programmable weight matrix of the form

$$W = \begin{matrix} W_{i-1,j-1} & W_{i-1,j} & W_{i-1,j+1} \\ W_{i,j-1} & W_{i,j} & W_{i,j+1} \\ W_{i+1,j-1} & W_{i+1,j} & W_{i+1,j+1} \end{matrix}$$

The kernel can be of any desired size, and the weights W_{ij} and so forth can have any form but are typically real and often integers. The simplest example would be when all weights are unity, resulting in a local mean. In this case, the operation starts with the original image, I, being stored in some memory plane of the machine and a copy of this being passed bit-serially, using an individual bus line for each pixel, to another memory plane in the machine.

The original image is then successively shifted north, east, south, and west to cover all positions in the 3×3 kernel. At each shift position a copy is sent, again bit-serially and one pixel for each bus line, to an accumulator, as shown in Fig. 8, to form the final summation. The full operation takes a total of eight shifts and additions or equivalent, to about 15 μs. For arbitrary weighting schemes an identical procedure is employed, but in these cases parallel multiplications are performed in the accumulator plane after each shift. Extensions of this basic concept can be used to form arbitrary local area operations for edge extraction, filtering, and feature extraction, all of which are important at the low level of image analysis.

4.2 Local histogramming operations

Other functions, such as histogram and median filtering, map very elegantly onto the 3-D structure. The technique for developing the histogram is illustrated in concept in Fig. 9. At the start of this operation, in addition to storing the original image on one memory plane, a template of nonbinary values ranging from zero intensity to the maximum is stored in uniform columns across the wafer as shown. Two other planes are also involved: a comparator and an accumulator. The principle is then to apply the image pixel value and the fixed template value as inputs to the comparator plane. If this is done and the *equal to* response interrogated as the image data are moved step by step relative to the template, an output will be generated and fed to the accumulator whenever the template value and pixel intensity match.

The result of this is a distribution of the pixel intensity values stored in columns across the accumulator plane. At this point, all that will be required to form the complete histogram is a stepwise accumulation down all the columns as illustrated. The overall processing time required for this proce-

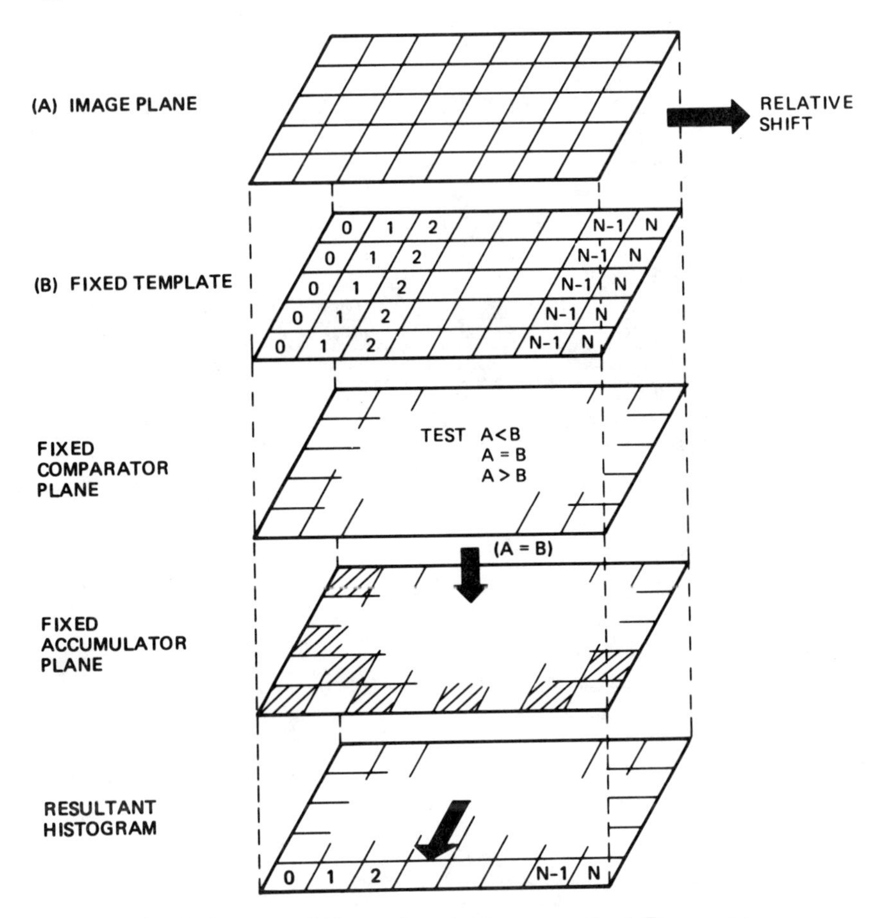

Fig. 9 Concept of histogram calculation on the 3-D computer.

dure is approximately 1.3 ms. More details of this procedure are given in Tsaur *et al.* [8], but it should be noted that in some cases, when the physical computer array size is less than the dynamic range of the image, the template might have to be distributed over several wafers. This causes a minor complication in that the operation proceeds in multiple steps, one for each template, but the basic concepts remain the same.

4.3 Median filtering operations

The histogram concept also allows additional capabilities to be exploited in performing nonlinear filtering operations such as median filtering. This

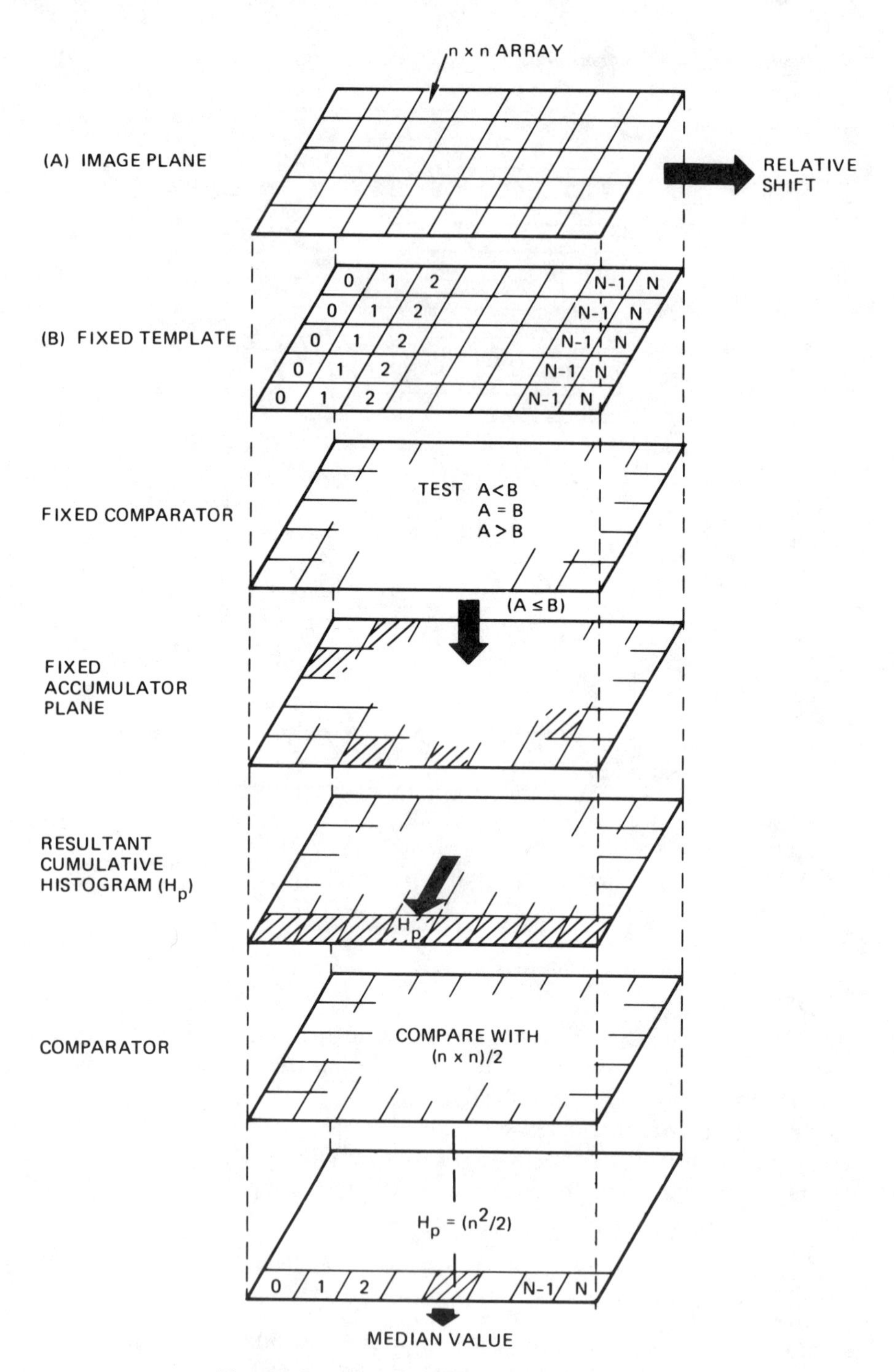

Fig. 10 Concept of parallel median calculation

operation is frequently used to remove impulsive noise without the resulting lack of resolution created by local averaging. Also, it can be used as a size filter in certain discrimination operations [9]. The concept requires that for a given $N \times N$ kernel size (32×32, for example) the central pixel value be replaced with a value equivalent to the median of all the interior values. This is typically a complex sorting operation and, for reference, might take several hours on a serial machine with the throughput available today. Our technique proceeds exactly as for the histogram. All pixels are again compared in parallel with the fixed template as the image data are shifted step by step across the water. This time the data accumulation is performed by selecting the *less than or equal* output from the comparator. This has the effect, after summation down the columns, of providing the cumulative histogram. At this point the midpoint of the histogram is found by an in-place parallel comparison using the histogram distribution and the value of $N \times N/2$ (512) for a $N \times N$ kernel. The resulting pixel intensity is the median. A schematic of the operation is shown in Fig. 10. The execution time is of the order of 35 ms.

5. SUMMARY

The examples given above on the use of the 3-D processor for image analysis are in no way intended to be an exhaustive list. Our intent is rather to provide an insight into the processing philosophy. Two issues are important in appreciating the power of the machine. First, with a cellular SIMD machine of this type, substantial throughput increases are obtained by the word-parallelism. Hence, though we are calculating the local $N \times N$ mean, for example, on one pixel site, it is automatically achieved for all points in the array. This parallelism, which is of the order of 10^3 in our initial demonstration and will increase to 10^5 in our ultimate machine, far outweighs the disadvantage of bit-serial arithmetic. Second, a principal disadvantage of the SIMD cellular architectures developed with conventional technology is the problem associated with memory access and loading the array. This situation is circumvented to a large extent in our processor by having memory planes (potentially many copies of them) distributed throughout the stack and by having a highly parallel bus structure, one for each cell, through which the data are transmitted. This implies that a full frame of data can be transmitted in approximately 2 μs. The additional advantages in terms of microelectronic component technology, device packing, power dissipation, and increased gate speed can, we believe, provide the computing advances required for real-time operation.

As indicated in the text, several technological issues remain to be developed prior to achieving our goal of a 512 × 512 machine with more than ten operating wafers. The overall yield to achieve a processor with acceptable cost and the thermal issues as key factors in the successful development of the 3-D architectures. For this reason we are implementing our structure in a combination of CCD and CMOS technology. In this case the power dissipated is proportional to the product of the total number of active gates and the effective frequency. Our calculations show that the thermal problem will be equivalent to the current VHSIC II goal and, hence, will be manageable.

In a subsequent paper we shall describe the use of the machine for higher level operations including feature classification and relational processing. Further, our simulations and software developments to date have not been performed exclusively for image analysis, and we have found significant advantages for linear systems solution, matrix algebra, synthetic array radar, and other applications. The benefits in these applications indicate a simultaneous improvement in power, volume, and throughput of as much as one order of magnitude.

REFERENCES

[1] Duff, M. J. B., (1978). Review of the CLIP image processing systems. *Proc. National Computer Conf.*, pp. 1055–1060.

[2] Batcher, K. E., (1980). Design of a massively parallel processor. *IEEE Transactions on Computers*, **C-29**, No. 9, pp. 836–840, September.

[3] Nudd, G. R., Nash, J. G., Narayan, S. S., and Jain, A. K., (1983). An efficient VLSI structure for two-dimensional data processing. *IEEE International Conference on Computer Design: VLSI in Computers*, Port Chester, New York, October 31–November 3.

[4] Rosenberg, A. L., (1981). Three-dimensional integrated circuitry. In *VLSI Systems and Computations* (H. T. Kung, Bob Sproull, and Greg Steele, eds.). Computer Science Press, pp. 69–80.

[5] U. S. Patent Application Serial No. 342630. (1982). An array processor architecture utilizing modular electronic processors. Filing date January 26.

[6] Etchells, R. D., Grinberg J., and Nudd, G. R. (1981). Development of a three-dimensional circuit integration technology and computer architecture. *Society of Photographic and Instrumentation Engineers*, **282**, 64–72. Washington D. C., April.

[7] Grinberg, J., Nudd, G. R., and Etchells, R. D. (1984). A cellular VLSI architecture *IEEE Computer Magazine*, January, pp. 69–81.

[8] Tsaur, B-Y, Fan, J. C. C., Geis, M. W., Chapman, R. L., Brneck, S. R. J.,

Silversmith, D. J., and R. W. Mountain, (1982). Electrical characteristics and device applications of zone-melting-recrystallized Si films on SiO_2. *Laser-Solid Interactions and Transient Thermal Processing of Materials*, Vol. 13 (J. Narayan, W. L. Brown, and R. A. Lemons, eds.). Elsevier-Science Publishing Co., p. 593.

[9] Nudd, G. R. (1980). Image understanding architectures. *National Computer Conference, Anaheim, California, May 1980, AFIPS Conference Proceedings*, Vol. 49, pp. 377–390.

Chapter Twelve

Some Remarks Concerning the Use of Superlattices as Shift Registers

J. D. Becker

1. INTRODUCTION

As the number of processing elements in a parallel processing machine increases, some kind of organization becomes necessary in all but some trivial cases. Hierarchical structures are the most obvious and familiar forms of organization, even if they might not be the most efficient ones (cf. e.g., Vester [1]).

An example for an hierarchically organized machine was given by Fritsch [2]: each processing element of a hierarchical level controls four processing elements of the next level below.

Since the ratio of the population n_h of a given hierarchical level h to the population n_{h+1} of the next level $h + 1$ is fixed,

$$n_h/n_{h+1} = 1/4$$

it is tempting to assume that such a machine is a hierarchical modular system (HMS) in the sense of Caianiello [3]. In this theory, a value v_h is assigned to each level h, and by definition the ratio

$$v_h/v_{h+1} = M = \text{const}$$

is independent of h. One of the consequences of the theory of HMS is a connection between this modulus M and the populations of the levels.

INTEGRATED TECHNOLOGY FOR
PARALLEL IMAGE PROCESSING
ISBN 0-12-444820-8

We find

$$n_h/n_{h+1} = M^{-1/2}$$

so that in our example the modulus is 16.

It is surprising that the value increases by such a large factor from one level to the next, but we shall not discuss this problem. We shall rather ask what determines the value of a processing element. We are not able here to answer quantitatively, but certainly the value will be a monotonically increasing function of the ability, hence of the knowledge, and hence of the size of the local memory of a processing element.

The lack of local memory is already one of the problems of parallel machines. Our considerations suggest that tomorrow this problem will grow faster than the number of processing elements in parallel machines.

It is unlikely that planar VISI technology will be able in the future to cope with the needs of parallel processing [4], [5]. An interesting attempt to conquer the third dimension is the wafer stack approach (see Chapter 11). In the long run, Molecular Beam Epitaxy (MBE) should be a suitable technology for 3-D structures. We shall describe this technology in Section 2.

MBE, however, also offers the possibility of designing novel elements. As a first example we consider superlattices, which are artificial periodic structures in the semiconductor crystal with a period of 50–1000 Å. The physics of such superlattices is briefly described in Section 3.

In Section 4 we shall discuss the possibility of using superlattices as shift registers. This would permit us to store large amounts of information in small volumes of the crystal.

2. MOLECULAR BEAM EPITAXY (MBE)

MBE is the 2-D monocrystalline growth of arbitrarily doped layers from the gaseous phase on a monocrystalline substrate, the wafer. In contrast to other epitaxial growth techniques, MBE works far from thermal equilibrium: The substrate is much cooler than the vapor.

The prerequisites are

(a) an ultrahigh vacuum (rest gas pressure $\leqslant 10^{-10}$ bar);

(b) a perfectly clean surface of the substrate; in particular, no trace of C or O;

(c) a substrate temperature of 400–750°C (for Si);

(d) an electron beam evaporator (for Si) and effusion cells for the doping materials (e.g., Sb and Ga).

Rates are controlled by a mass spectrometer. (The crystalline growth can be monitored by Reflective High Energy Electron Diffraction.) Figure 1 shows a schematic of an MBE plant for silicon epitaxy.

The important features of this technique are

(a) monocrystalline growth of arbitrarily doped layers,
(b) a wide range for the thickness of the individual layers,
(c) excellent control and reproducibility of doping and thickness,
(d) no diffusion of the doping materials because of the low temperature of the substrate.

For recent developments we refer to the literature [6], [7], [8]. Applications, so far, concentrate on devices for which thickness control is vital (e.g., transit time and tunneling devices) or on heterostructures such as GaAs/GaAlAs structures (DH lasers, HEMT transistors).

For application to parallel processing, MBE has to be combined with a method allowing for lateral structures. One such method would be to "write" with ionized beams of doping materials. Recently, effusion cells for ionized beams have been developed, because they allow for a fast rate control and because ions have a better sticking coefficient on the substrate [9], [10]. However, to write with an ion beam, a precise focus is necessary; thus, one would have to use beams of higher energy which would damage the crystal.

Another method would be to use effusion cells with a fast rate control by continuously variable shutters [11] combined with a masking technique that has yet to be developed. The main problem will probably be to clean the Si

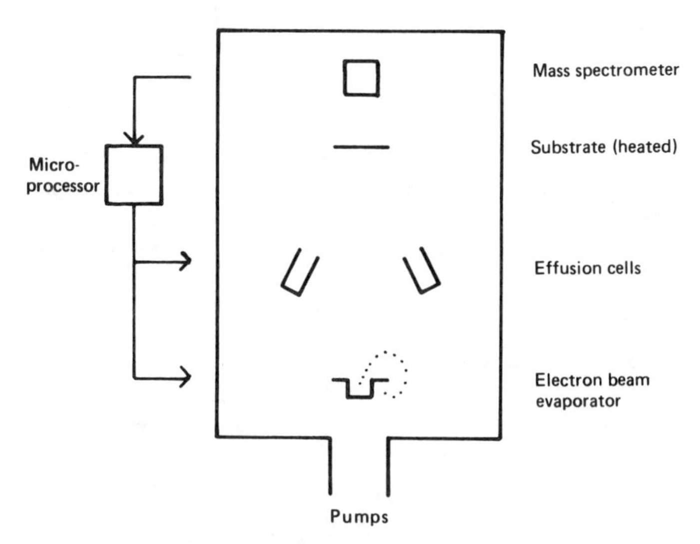

Fig. 1 MBE plant for silicon epitaxy, schematic

surface after masking; MBE growth will work only on a perfectly clean surface.

The most elegant way, however, would be an in situ lateral structuring; laser beam etching and other advanced technologies may help us with this problem.

3. SUPERLATTICES

Superlattices are monocrystalline samples with an additional periodic structure that typically has a period of 50–1000 Å. So far only 1-D superlattices have been made, with the *superstructure* in the *direction of epitaxial growth*; we shall refer to this direction as the *z direction*.

There are two ways of making superlattices (Fig. 2): by alternative deposition of different materials with a similar lattice constant (hetero-superlattices) and by a periodic variation in the doping concentration (doping superlattices). (In principle, one might combine the two methods.)

In both cases, (as shown in Fig. 3,) the upper edge of the valence band and the lower edge of the conduction band become modulated with the period of the superlattice [12], [13].

In heterostructure superlattices, the chemical potential (i.e., the Fermi level) must be constant throughout the structure (while in equilibrium). Since in each material the band gap is fixed, the band edges arrange themselves correspondingly. Usually one gets a contravariant modulation of the band edges. In doping superlattices the periodic modulation is caused by space charges (ionized donor and acceptor atoms). In this case one obtains a

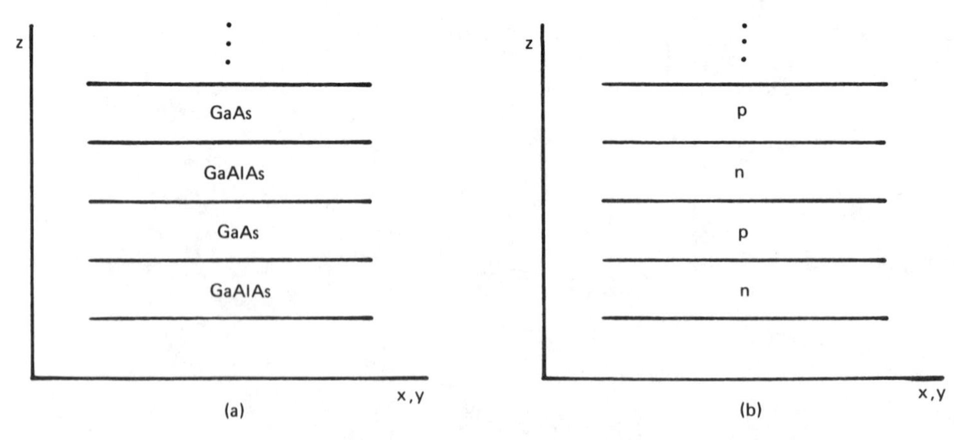

Fig. 2(a) Heterostructure superlattice and (b) doping superlattice

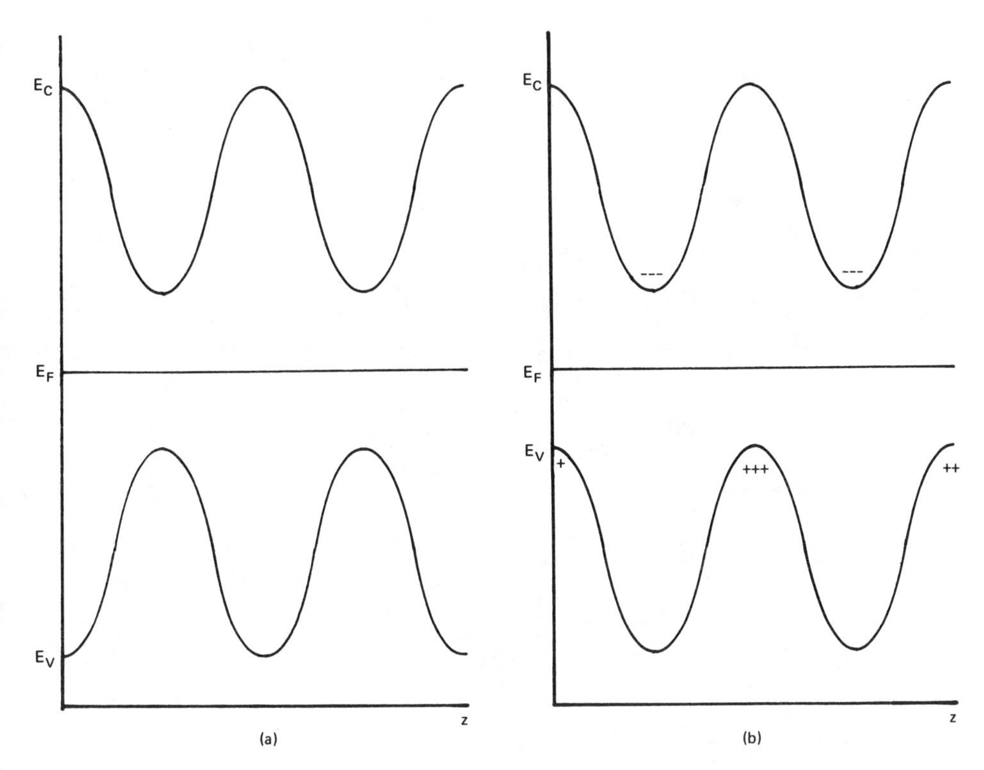

Fig. 3 Energy band edge modulation in a superlattice: (a) heterostructure and (b) doping superlattice

covariant modulation of the band edges. In compensated superlattices (i.e., on the average the donor and acceptor densities are equal) things are particularly simple: Most of the free electrons and holes recombine, the band edge modulation is simply the potential caused by the dopants, and the Fermi level lies practically in the middle between the bands.

The shape of the modulation may be varied by appropriately choosing the doping profiles.

Doping superlattices show many interesting features [14]. For our purpose the most interesting property is the spatial separation of electrons and holes, as indicated in Fig. 3, which results in a greatly enhanced lifetime (by many orders of magnitude) of the charge carriers and hence of nonequilibrium states.

If the valleys in the modulation of the band edges are narrow and deep enough, quantized levels will appear. In a compensated superlattice in the ground state, it is easy to calculate these levels from the shape of the band modulation taken as the potential in the Schrödinger equation. In non-compensated superlattices, as well as in highly excited superlattices, the calculation of these levels (together with the exact modulation) is a complicated problem of self-consistency. If we work in the effective-mass approximation, we may factor the Schrödinger equation. In the planes orthogonal to the z direction we can assume plane waves because the extension of the superlattices in x- and y-directions is much greater than in the z-direction. For the z-direction, we must solve self-consistently the following set of equations (low T approximation):

(a) the 1-D Schrödinger equation,

$$\left[-\frac{\hbar^2}{2m_z^*}\frac{d^2}{dz^2} + V(z)\right]\psi_{sk}(z) = E_s(k)\,\psi_{sk}(z)$$

(b) the normalization equation to determine the Fermi energy E_F,

$$N = \frac{m_z^* d}{\pi \hbar^2}\sum_s \int_{-\pi/d}^{\pi/d} \frac{dk}{2\pi}[E_F - E_s(k)]\,\theta\,[E_F - E_s(k)]$$

(c) the equation for the electron density,

$$\rho(z) = \frac{m_z^* d}{\pi \hbar^2}\sum_s \int_{-\pi/d}^{\pi/d} \frac{dk}{2\pi}[E_F - E_s(k)]\,\theta\,[E_F - E_s(k)]/\psi_{sk}(z)/^2$$

(d) the Poisson equation,

$$\frac{d^2V}{dz^2} = -\frac{e^2}{\epsilon}[\rho(z) - N_D(z) + N_A(z)]$$

where s = level index, N = areal density of electrons, d = superlattice period, ϵ = dielectric constant and N_D and N_A = concentration of donors and acceptors. The total energy is then given by

$$E_{\text{tot}}(\vec{k}) = E_s(k_z) + \frac{\hbar k_x^2}{2m_x^*} + \frac{\hbar k_y^2}{2m_y^*}$$

Such computations have been carried out for superlattices by Ando and Mori [15], Zeller *et al.* [16], and Ruden and Döhler [17]. In principle, however, there is no difference between their results and the results obtained in the simple case of a compensated superlattice in the ground state.

The low-lying energy levels of the states in the superlattice valleys have virtually no k_z dependence, i.e., no level widths, which means that the electrons are well localized in the individual valleys. Electrons occupying the

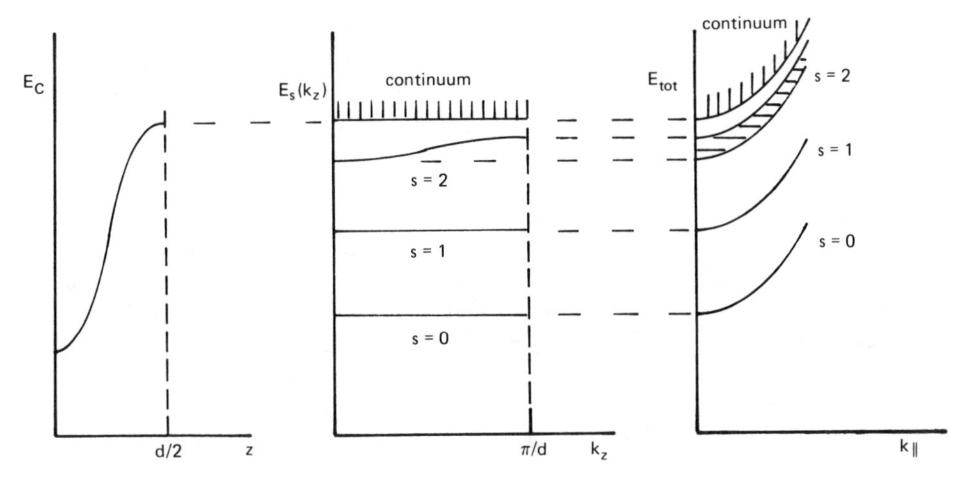

Fig. 4 Dispersion of minibands in a superlattice

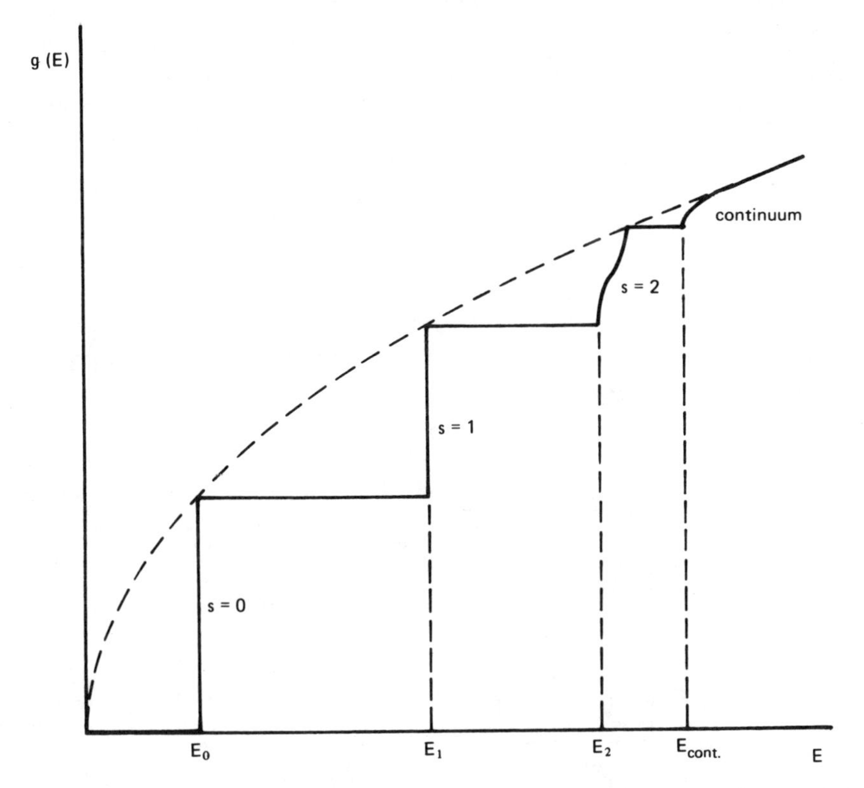

Fig. 5 Density of states in a superlattice (dashed curve: density of states in the bulk)

higher states can tunnel through the superlattice more easily, so these levels develop some k_z dependence and are thus called *minibands* (Fig. 4).

Since in two dimensions the density of states is independent of E, $g(E)$ is constant between the minibands and changes rapidly within them (Fig. 5).

4. THE SUPERLATTICE SHIFT REGISTER

One bit of information in VLSI technology is represented as the presence or absence of a charge package of $\sim 5 \times 10^4$ elementary charges (corresponding to an areal density of 5×10^{12} cm^{-2} spread over an area of 1 μm $\times$ 1 μm). An elegant way of shifting such charge packages around is realized in the Charge Coupled Device (CCD). In the CCD, the charge packages are stored in small capacitances under the insulated electrodes. They can be shifted by appropriate changes in the voltages. Three different voltages are required. The CCD needs periodic refreshing.

A superlattice is also a natural structure to store charge packages; the only question is how to shift them. It is unlikely that it will be possible to make gates for switching individual layers of the superlattice. Hence, a dynamic mechanism is required to shift the charge packages. We suggest a mechanism combining *resonant tunneling* between neighbouring valleys and *thermal relaxation* in the valleys in a two-phase step (*sample and hold*) that is controlled by voltages applied externally to the whole superlattice shift register. We shall now discuss the shift procedure in a simplified manner.

One cell of the shift register consists of two valleys that are not identical, each of which contains two quantum levels, as shown in Fig. 6.

The upper levels are thermally broadened and have larger tails extending further into the hills, whereas the lower levels are sharp, and the corresponding states are well localized.

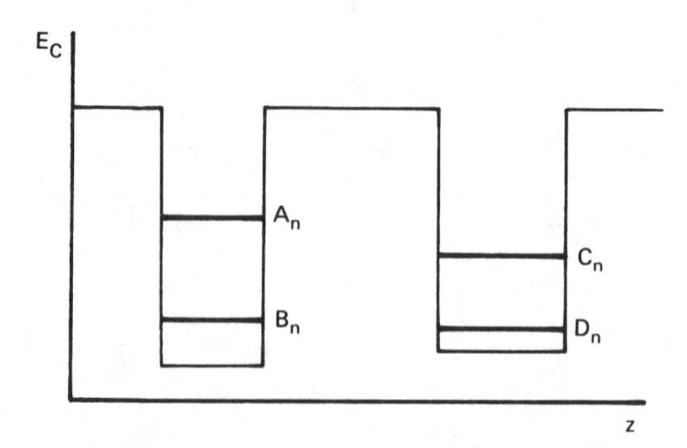

Fig. 6 Superlattice shift register cell #*n* (qualitatively)

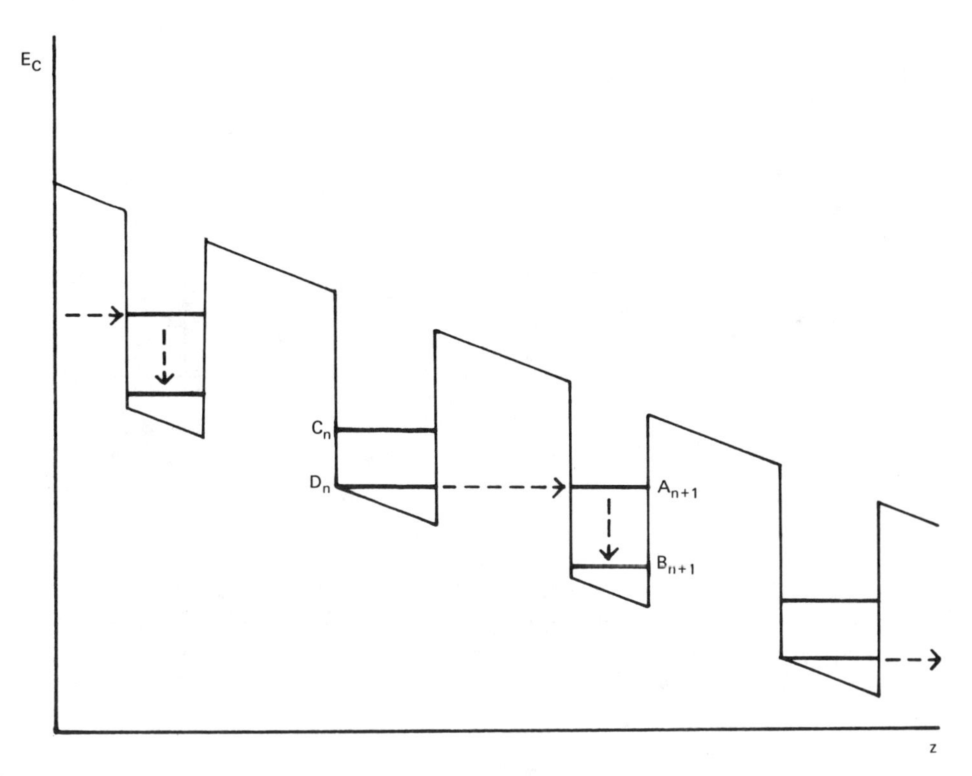

Fig. 7 Sample step (fast, field high) (qualitatively)

The charge packages are shifted into the $+z$ direction. The left valley is used for temporary storage in the sample step; the right valley will contain the charge in the hold step.

For the sample step an electric field is applied to the superlattice so that level D_n of cell #n is in resonance with level A_{n+1} of the neighbouring cell #$n + 1$, as in Fig. 7. If D_n contains a charge package, it will tunnel to the level A_{n+1} and thermalize into the level B_{n+1}, which is a metastable state.

For the hold step the external field is changed so that the levels B_{n+1} and C_{n+1} are in resonance; if B_{n+1} contains a charge package, it will tunnel into C_{n+1} and thermalize into D_{n+1} (see Fig. 8). Rough estimates support the idea that it should be physically and technologically possible to make such shift registers in practice.

Shift registers would consist of *towers* grown on the wafer that

(a) have an area of 5×5 μm^2,

(b) have a height of 8 μm,

(c) contain 100 cells of a height of 800 Å each.

The *doping concentrations* would be 3×10^{18} cm^{-3}.

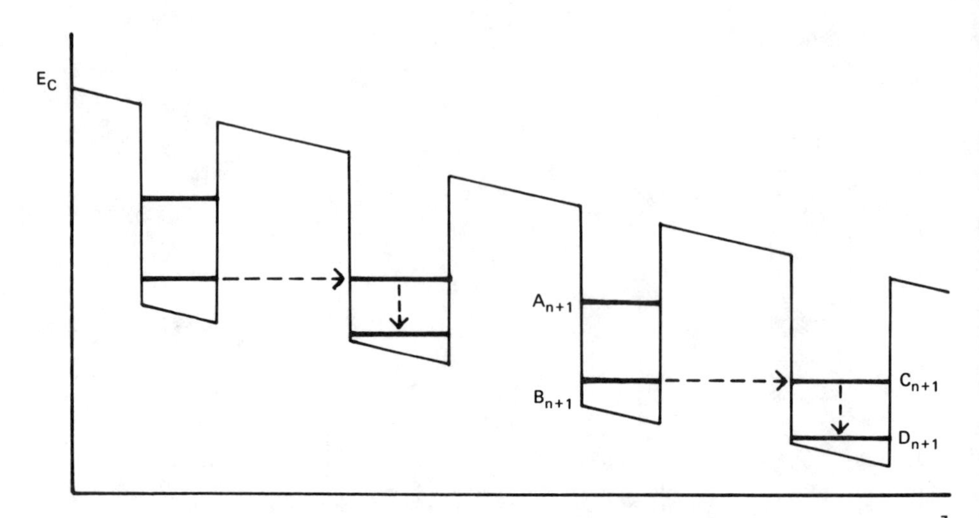

Fig. 8 Hold step (slow, field low) (qualitatively)

Each *cell* can contain a *charge* package of 5×10^4 electrons $= 8 \times 10^{-15}$ Cb, corresponding to an *areal density* of $\sim 2 \times 10^{11}$ cm^{-2}.

The operating *temperature* would be 77 K, i.e., at the liquid nitrogen boiling point (the liquid nitrogen will also cool the chip).

The *energy* scales are given by

(a) the barrier height of the modulated conduction band edge, ~ 100 meV;

(b) the energy level differences, which are of the order of 30 meV;

(c) the thermal width of the upper levels, $kT \sim 7$ meV at 77 K.

The current *voltage* supply to chips, 5 V, is sufficient to drive the shift register; at each cell we would have a voltage of $\leqslant$50 mV.

We now list the various *time* scales. The shift register could operate with

(a) a shift cycle time of 10^{-6} s,

(b) a sample-and-hold cycle time of 10^{-8} s,

(c) a switching time of 10^{-9} s,

(d) a charge transfer time of 10^{-11} s (see below),

(e) a thermal relaxation time of $\hbar/kT \sim 10^{-13}$ s,

(f) a tunneling time of $\hbar/(V - E) \sim 10^{-14}$ s.

The most difficult estimate is for the charge transfer time; however, it can easily be adjusted by an appropriate choice of the barrier, because the tunneling probability depends almost exponentially on that width. The characteristic width is

$$\kappa^{-1} = \frac{\hbar}{[2m^*(V - E)]^{1/2}} \sim 50-100 \text{ Å}.$$

If we use the semiclassical formula for the transfer time,

$$\tau = \frac{2 \times \text{valley width}}{\text{classical velocity}} \times (\text{tunneling probability})^{-1} = \frac{2a}{(2E/m^*)^{1/2}\ T^*T}$$

we arrive at the conclusion that a charge transfer time of 10^{-11} s can be achieved with a *barrier width* of ~300 Å.

Last, let us consider the *power loss*. For one tower, it is given by

$$P = U \times I \leqslant U \times \frac{\#\text{cells} \times \text{charge per cell}}{\text{shift cycle time}} \sim 5 \times 10^{-6}\ \text{W}$$

Thus, if we assume a total number of 10,000 shift register towers on one chip, each containing 100 bits, we would arrive at a total power dissipation of 50 mW, which is quite acceptable.

5. DISCUSSION AND CONCLUSION

We have shown that, in principle, MBE-grown superlattices can serve as shift registers and that in a volume of $5 \times 5 \times 8\ (\mu\text{m})^3$ 100 bits of information can be stored.

The shift mechanism suggested here is based on resonant tunneling and thermal relaxation. It relies on the facts that levels in a superlattice are quantized; that lower states are well localized and have a sharp energy, whereas upper states are thermally broadened and more extended; that thermal relaxation prevents electrons from tunneling back; and that nonresonant tunneling probabilities are suppressed by a factor of $\sim 10^{-4}$ [18] and occur only during the short sample steps.

A number of questions remain. Among the physical problems, we find that additional charges change the potential (for the figures given above, the change should be smaller than the thermal width, however), that the influence of donor levels has been neglected; that electron–hole pairs will be generated because external voltages are applied, and that impurity and phonon scattering have been discarded. There are certainly a number of unwanted and disturbing processes, and the potential has still to be optimized to suppress these effects sufficiently. Some of these problems will be discussed in Becker [19].

We also encounter technical problems; for instance, traps may disturb the transfer of charge, and tunneling barriers may have a limited lifetime. However, it seems that these problems will not seriously effect the functioning of the Superlattice Shift Register.

Finally, let us make a somewhat indecent speculation. If doping superlattices can be used as shift registers for electrons, one may ask whether one could use holes to simultaneously transfer information in the opposite direction.

ACKNOWLEDGEMENTS

The author is indebted to Dr. G. H. Döhler for a clarifying discussion concerning the bandstructure of superlattices and to Mr. A. Beck, Prof. Dr. I. Eisele, and Dr. P. Vitanov for discussions concerning technological questions.

REFERENCES

[1] Vester, F. (1983). In *Parallel Processing: Organization and Technology* (J. Becker, B. Bullemer, and I. Eisele, eds.). Springer, New York.
[2] Fritsch, G. (1983). In *Parallel Processing: Organization and Technology* (J. Becker, B. Bullemer, and I. Eisele, eds.). Springer, New York.
[3] Caianiello, E. (1977). *Biol. Cybernetics* **26**, 151.
[4] Becker, J. and Eisele, I. (1984). In *Cybernetic Systems: Recognition, Learning, Self-Organisation* (E.R. Caianiello and G. Musso, eds.). Research Studies Press, Letchworth.
[5] Eisele, I. (1983). In *Parallel Processing: Organization and Technology* (J. Becker, B. Bullemer and I. Eisele, eds.). Springer, New York.
[6] Farrow. R. F. C. (1981). *J. Vac. Sci. Technol.* **19**, 150.
[7] Bean, J. C. (1981). In *Impurity doping processes in silicon* (F. F. Y. Wang, ed.). North-Holland Publ., Amsterdam.
[8] Foxon, C. T., and Joyce, B. A. (1981). In *Current Topics in Material Science*, Vol. 7 (Kaldis, E. ed.) North-Holland Publ., Amsterdam.
[9] Naganuma, M., and Takahashi, K. (1975). *Appl. Phys. Lett.* **27**, 342.
[10] Matsunaga, N., Suzuki, T., and Takahashi, K. (1978). *J. Appl. Phys.* **49**, 5710.
[11] Beck, A., Becker, J., Bullemer, B., and Eisele, I. (1981). *European Workshop on Molecular Beam Epitaxy*, MPI, Stuttgart.
[12] Shik, A. Ya. (1975). *Sov. Phys. Semicond.* **8**, 1195.
[13] Ando, T., Fowler, A. B., and Stern, F. (1982). Rev. Mod. Phys. **54**, 437.
[14] Döhler, G. H. and Ploog, K. (1979). *Progr. Crystal Growth Charact.* **2**, 145.
[15] Ando, T., and Mori, S. (1979). *J. Phys. Soc. of Japan* **47**, 1518.
[16] Zeller, Ch., Vinter, B., Abstreiter, G., and Ploog, K. (1982). *Phys. Rev.* **B26**, 2124.
[17] Ruden, R., and Döhler, G. H. (1983). *Phys. Rev.* **B27**, 3538.
[18] Tsu, R., and Esaki, L. (1973). *Appl. Phys. Lett.* **22**, 562.
[19] Becker, J. (1983). In *Parallel Processing: Organization and Technology* (J., Becker, B., Bullemer, and I. Eisele, eds.). Springer, New York.

Chapter Thirteen

Plans for the CLIP7 Chip

T. J. Fountain

1. INTRODUCTION

Two previous volumes in this series, derived from the workshops held at Ischia in 1980 [1] and at Abingdon in 1982 [2], contained chapters about the CLIP4 bit-serial image processor chip and the increasing range of fundamentally similar devices proposed by other workers. This chapter presents the design and plans for use of a new device—the CLIP7 chip—which differs in a number of ways from the type of processor described in Duff [2], although its intended first use in arrays of processors is similar. Some of these changes result from a recognition that types of architecture and control strategy other than those employed in the SIMD type of array processor offer the potential for improved performance and new insights in the field of image processing. However, it must also be recognised that the more complex the system, the more difficult it is to devise worthwhile algorithms and control strategies. Our intention, using the CLIP7 chip as a basis, is to employ an evolutionary approach to system design and, hopefully thereby, to arrive at a clearer understanding of, and perhaps solutions to, the problems involved.

The goals of our program are threefold. First, the provision of a system for use in application studies, having substantially improved performance over CLIP4, in particular in pixel resolution. Second, the investigation of the behaviour of arrays of processors in which each individual has an increasing degree of local autonomy. Third, the construction of assemblies of processors having connectivities other than those used in the typical 2-D array processor. Structures of this type that are already proposed include pyramids of various types [3], [4], 3-D processors [5], and a number of bus-based systems, usually utilising microprocessors to provide computing power [6], [7].

INTEGRATED TECHNOLOGY FOR
PARALLEL IMAGE PROCESSING
ISBN 0-12-444820-8

The CLIP7 chip has been designed to fulfil all these requirements. The evolution of the design, its final structure, and its basic operations are described in the next section.

2. THE CLIP7 CHIP

2.1 Evolution

The CLIP4 chip and its many close cousins (see Duff [2]) all have a similar basic structure, shown in Fig. 1. Such bit-serial circuits are extremely flexible but lack individual power when compared with, for instance, a microprocessor. In addition, many of the most time-consuming operations in image processing are carried out on grey-level rather than binary data, suggesting that a multibit circuit might be an appropriate structure.

Initial studies of a variety of possible processor configurations [8], [9] led to two important conclusions. First, for circuits involving dispersed interconnection paths, a full-custom LSI design was more efficient than any other implementation, and in particular, it was more efficient than systems based on currently available gate arrays. Second, within the context of custom LSI, the amount of processing power available from a given area of silicon was substantially independent of the wordlength employed. In particular, it was possible to devise, within the same pin-count and number of devices, chips having respectively 16 single-bit processors and one 16-bit processor whose typical performances on currently used mixtures of algorithms would be very similar.

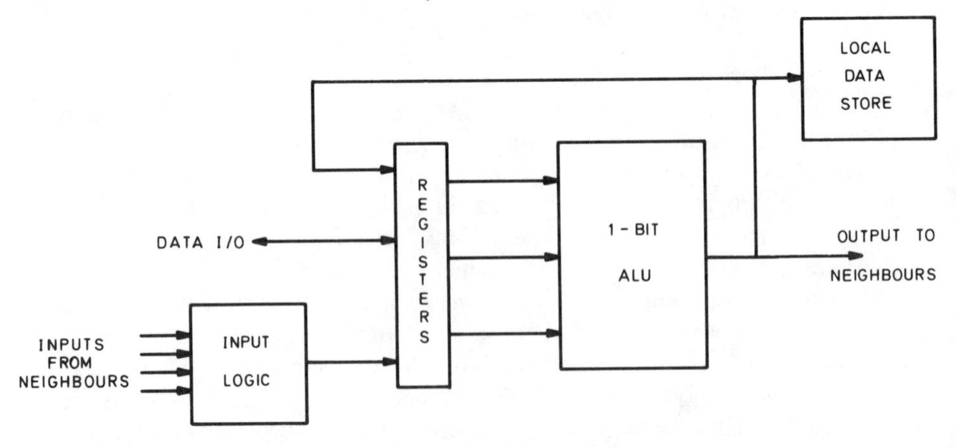

Fig. 1 Typical bit-serial processor structure

However, further studies indicated that the 16-bit configuration offered the possibility of incorporating additional architectural features of interest, and so this design was pursued to become the CLIP7 chip, described below.

2.2 Technology

One of the first decisions to be made in the design of a custom LSI chip concerns fabrication technology. At this early stage a schism develops between two groups of designers. One alternative is to utilise the most powerful technology available. The resultant device should be both architecturally and technologically advanced and should, therefore, represent state of the art in all respects. The other alternative is to use the current bread-and-butter technology, depending for the power of the device solely on architectural innovations and hoping to reap benefits of improved reliability and ease of design from the mature technology.

A choice between these alternatives is usually made on the grounds of available resources. Given equal expertise, it will often take many redesign cycles to perfect a high-technology device, whereas a mature technology will often yield a satisfactory result from the first design. Design strategies, therefore, usually split between (1) manufacturers' in-house design teams that have access to substantial technological expertise, backup services, and often a dedicated prototype processing line offering fast turn-around for redesigns and (2) buyers of technology, such as universities, which often work in tandem with independent design groups.

Our own decision was to design in a mature technology. In Britain, there are currently only two candidates in this category: NMOS, in either metal- or silicon-gate versions, and iso-planar CMOS. CMOS offers important benefits in power consumption and speed and was therefore chosen for the CLIP7 chip.

2.3 Structure

The logical structure of the CLIP7 chip is shown in Fig. 2, which depicts the principal data paths, and Fig. 3, which indicates how control of the chip is achieved. In general, data buses in the chip are 16 bits wide, whereas external data accesses are through 8-bit ports. The major exception to this scheme concerns the neighbourhood interconnections between processors. The original design concept employed 8-bit connections for this purpose, but packaging requirements forced the serialisation of these connections and,

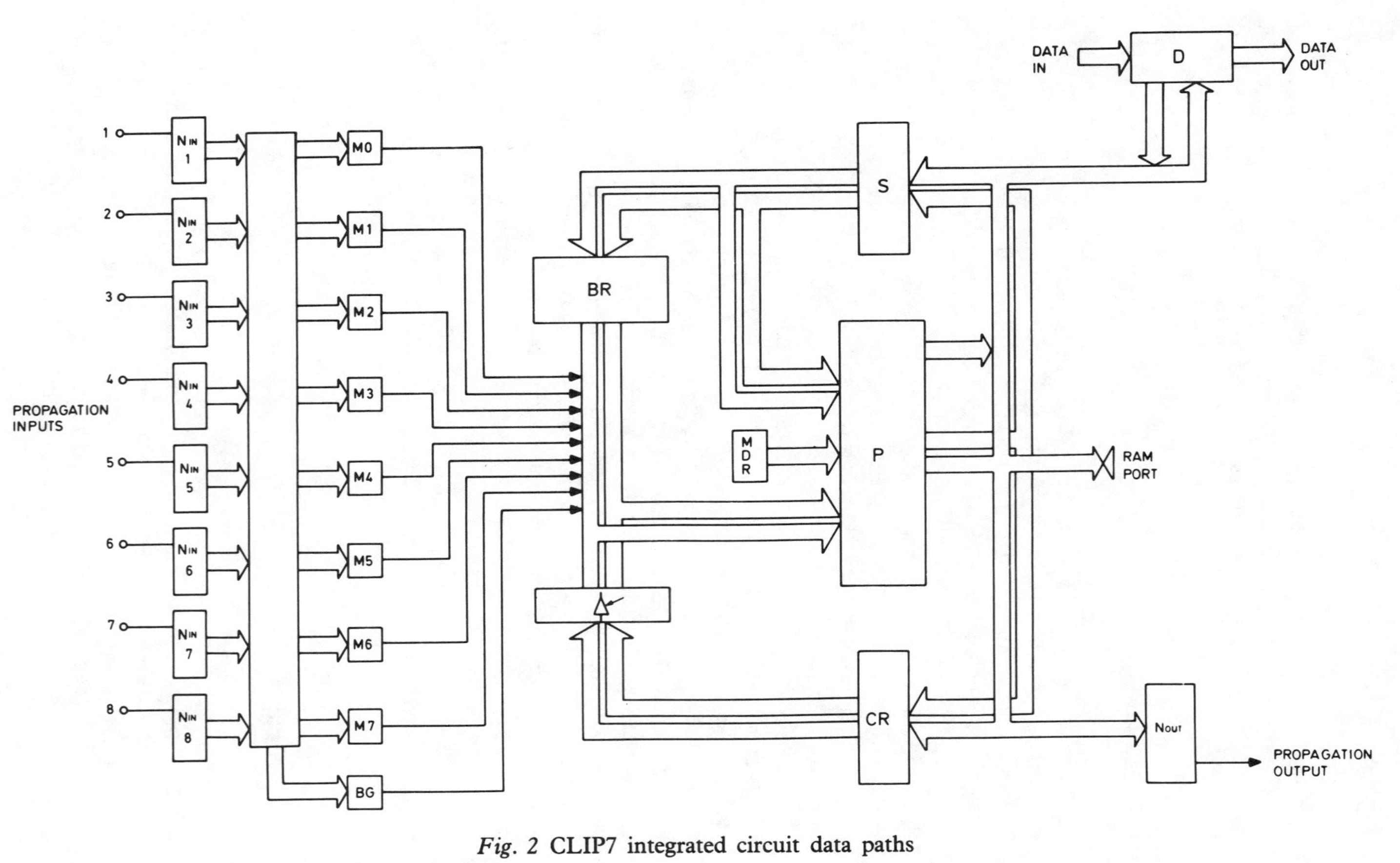

Fig. 2 CLIP7 integrated circuit data paths

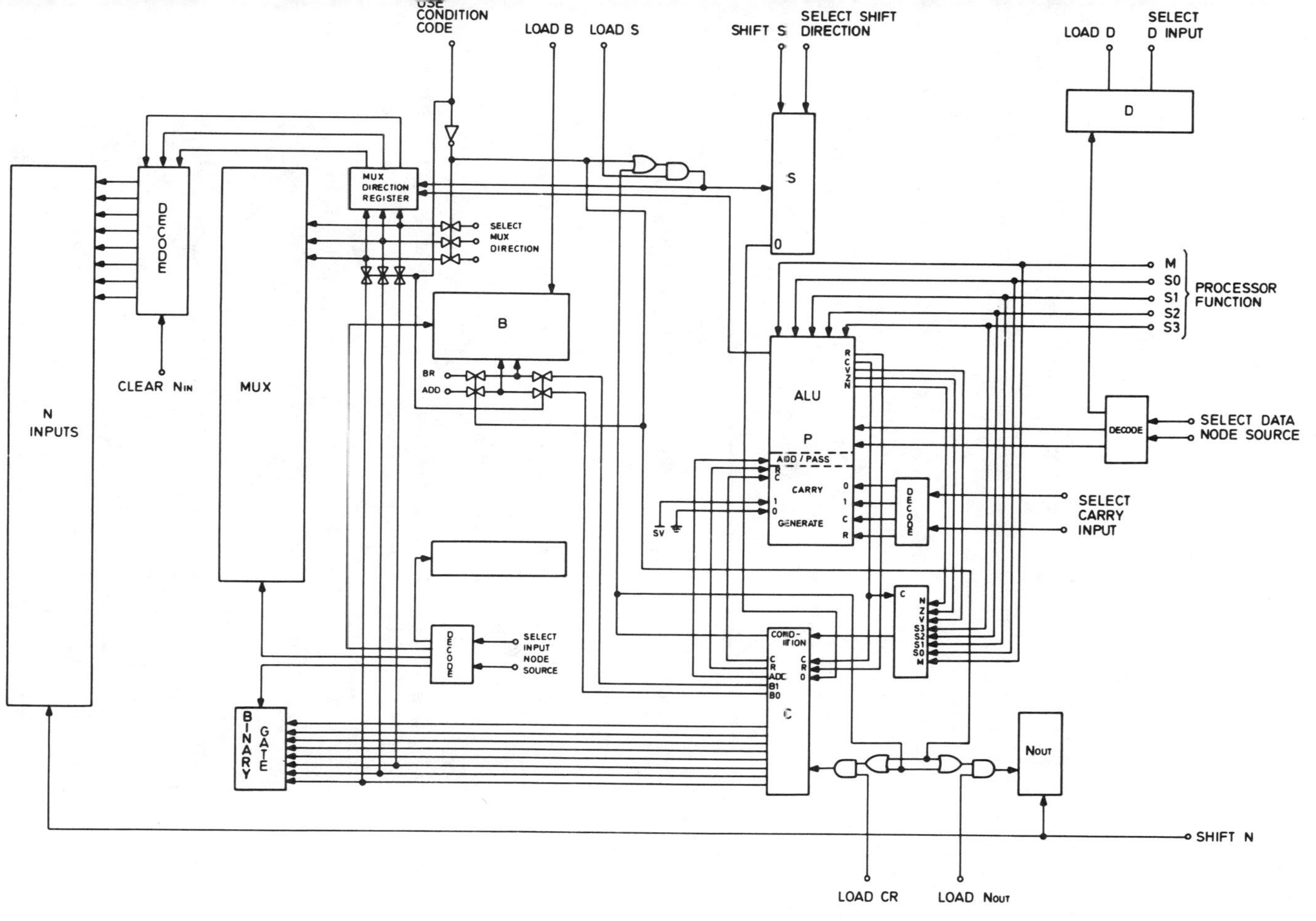

Fig. 3 CLIP7 chip control logic

Table 1
Bit assignments of the C register

Register bit	*Assignment*
0	Multiplexer direction select (bits 0–2)
1	Multiplexer direction select (bits 0–2)
2	Multiplexer direction select (bits 0–2)
3	Binary gate direction field (bits 0–7)
4	Binary gate direction field (bits 0–7)
5	Binary gate direction field (bits 0–7)
6	Binary gate direction field (bits 0–7)
7	Binary gate direction field (bits 0–7)
8	B register address 0
9	B register address 1
10	S register bit 0
11	ALU bit 2 (round up)
12	ALU carry output
13	Not assigned
14	Not assigned
15	Conditional load

thereby, the use of single lines. The major functional elements of the circuit are as follows:

1. P is a 16-bit, two-input ALU offering the 16 logical boolean combinations of its inputs as well as add and subtract functions.

2. S is a 16-bit bidirectional shift register having logical shift, arithmetic shift, and logical rotate operations.

3. B is a block of four addressable 16-bit registers.

4. C is a 16-bit dual-purpose register. As well as being used for normal data storage, it is principally concerned with conditional operations of the circuit. Used in this mode the bit assignments of the register are those shown in Table 1. Its functions in this conditional mode are described later.

5. D is an 8-bit register used as one of the two principal data ports of the circuit. It is chiefly intended for use as shown in Fig. 4 when interfacing to serial data devices such as frame stores or TV cameras. The second data access point is the RAM port, an 8-bit port allowing access to the low byte of the OBUS and the high byte of the SBUS. It will usually be connected to byte-wide RAMs used as local data storage for the processor.

6. The structures intended to provide connections between neighbours in an array of processors comprise the Nout register (8-bit), the eight Nin registers (each 8-bit), a 64- to 8-line multiplexer M and a binary gate BG. The connections between Nout and Nin registers in an array are shown in Fig. 5 and are, as explained earlier, bit-serial in nature. The detailed configuration

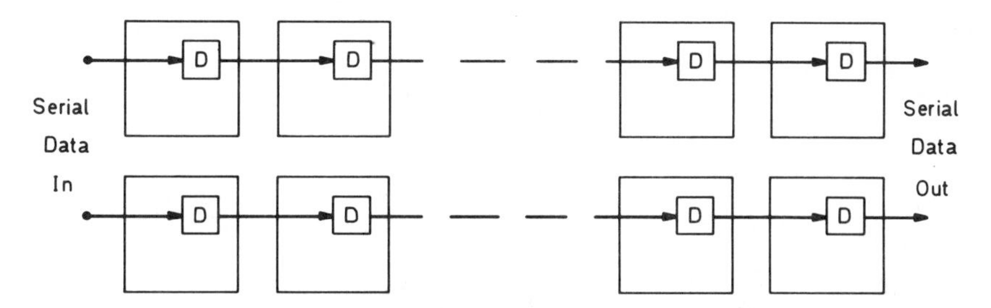

Fig. 4 Data I/O along rows of processors

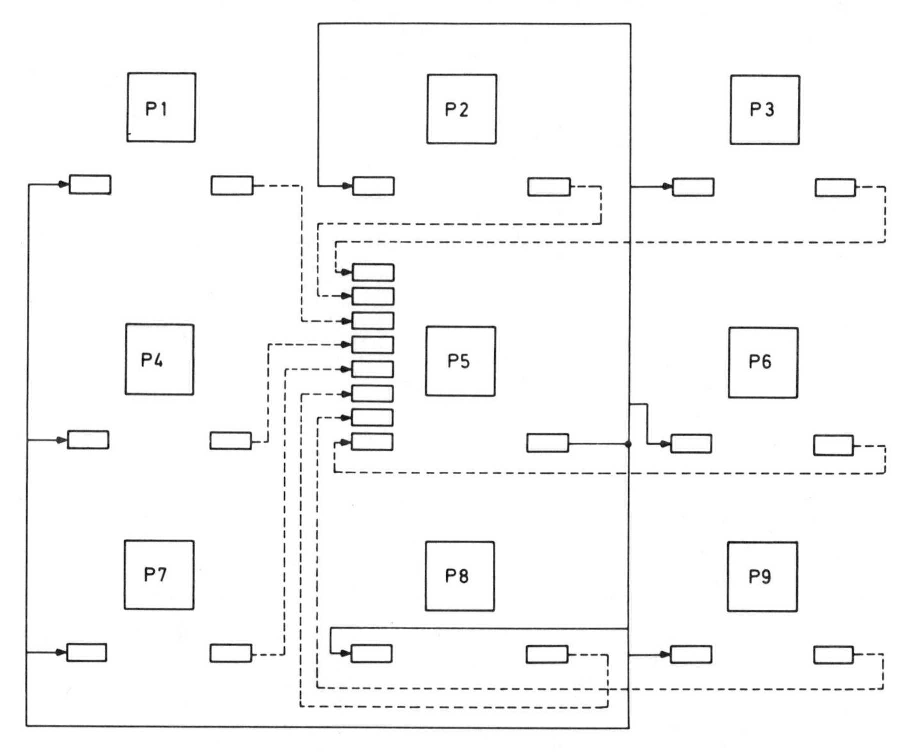

Fig. 5 Neighbourhood connections to and from P5

of Nin registers, M and BG, is shown in Fig. 6. This structure allows both 8-bit and binary neighbourhood values to be dealt with efficiently.

7. MDR is a 3-bit register that is loaded with the direction in which the multiplexer was pointing at the time the S register was loaded. Its outputs are passed, under certain circumstances, to the OBUS (bits 0–2). The use of this register in conditional processor operations will be explained below.

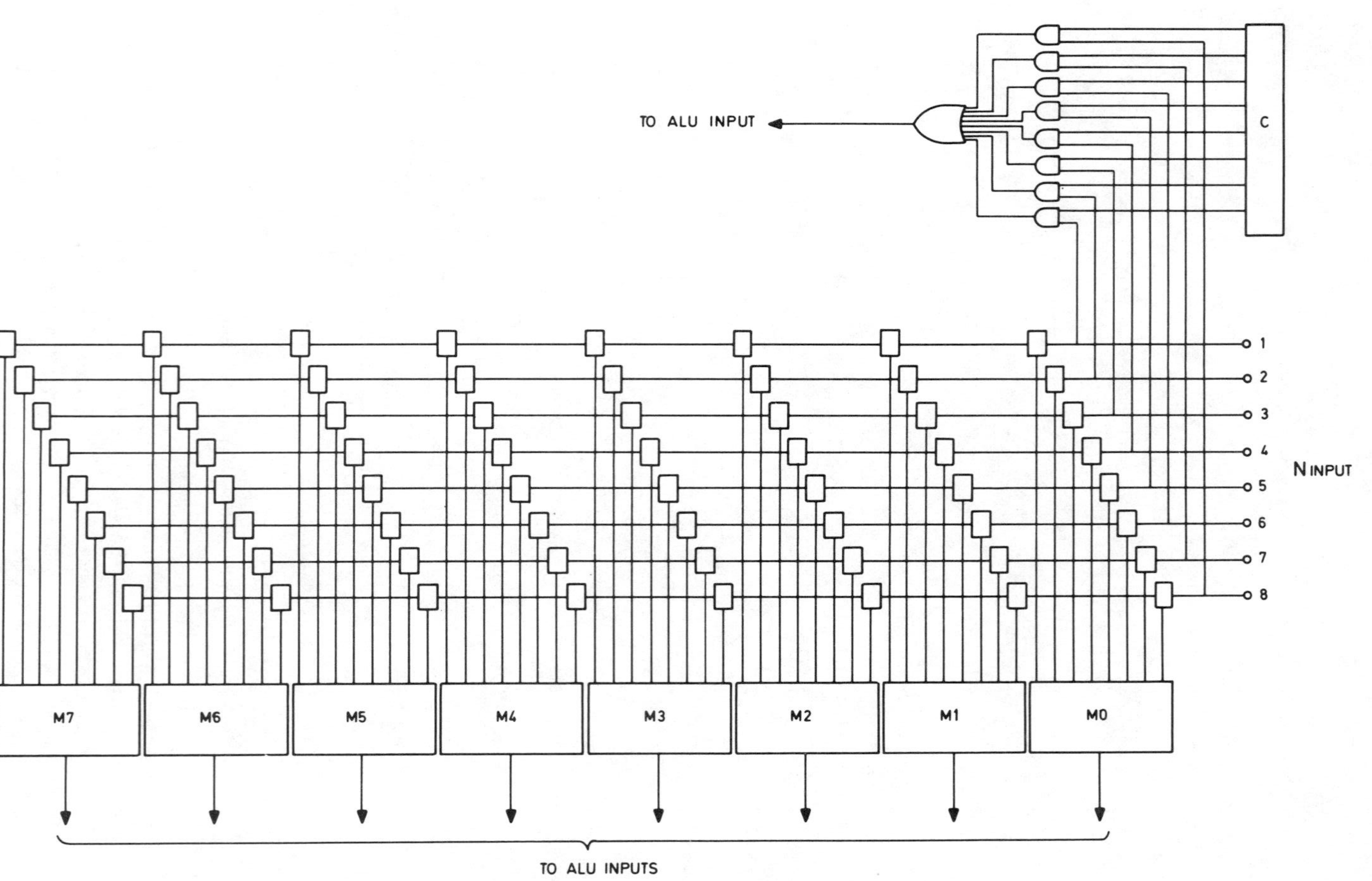

Fig. 6 CLIP7 chip input matrix

2.4 Novel features

The structure chosen for the CLIP7 chip incorporates a number of novel features that are worth emphasising at this stage, together with their implications.

First, the 16-bit architecture produces a processor of great power, approximately 200 times that of the CLIP4B processor. This means that SIMD processors of extreme speed could be constructed from such devices. Furthermore, the performance is achieved by architectural features, the chosen technology limiting clock rate to 5 MHz. The same architecture in a more powerful technology could yield further substantial increases in power with no conceptual rethinking required.

Second, the incorporation of the C and MDR registers in the chip data paths will allow the construction of systems in which each processor in an array has a degree of local autonomy. Normally, control of chip operation is by means of externally applied signals. However, if conditional operation is selected, the contents of the C register can effect chip operation in the following ways:

1. The eight lines controlling the directions selected for the binary gate are always derived from the C register (C0–C7).
2. The directional control of the multiplexer can be derived from C0–C2.
3. The address of the B register can be derived from C8–C9.
4. The low-order bit of the S register can be stored in C10. It is intended for use in multiplication and division operations and can be used to determine whether the ALU performs an addition (or subtraction) or merely passes one operand to the output.
5. Two of the four alternative carry inputs to the ALU are provided by C11 and C12. C11 is derived from the ALU output (bit 2) and is intended for use in rounding operations, whereas C12 is the previous ALU carry output.
6. Bit 15 of the C register can be loaded with any of the four condition outputs of the ALU (carry, overflow, zero, and negative) and can then be used to control the conditional loading of the S, Nout, and C registers.

The MDR is used, typically, in neighbourhood ranking operations of the median filter type. It records the direction from which a selected value arrived at the cell, and this information can be transferred to the OBUS. Such a function can be used in route- or graph-searching algorithms to enable each processor in an array to receive data from an internally specified direction.

The third principal novelty of the CLIP7 chip derives from its implementation as a single processor on one chip. The result of this is that all data lines are available externally, in particular the neighbourhood interconnection lines. This, in turn, means that it is possible to build assemblies of processors, having various connectivities, with comparative ease. A chip such as CLIP4, where many internal interconnections are preconfigured, makes any such attempt extremely difficult.

2.5 Basic operations

The basic modes of operation of the circuit comprise the following:

1. *Serial data transfers.* These can be either input or output operations and occur via the D registers, connected as shown in Fig. 4 and interfaced to suitable frame stores.

2. *Local data storage and retrieval.* These paired operations involve transfers between the OBUS and external storage via the RAM port.

3. *Pointwise operations.* Here the intention is to produce a result that is a function of two or more inputs already resident in the local RAM. A simple example is the addition of two numbers. The first operand moves from RAM to S register to B register. The second operand then moves from RAM to S register. The addition is performed in the ALU, and the low-order byte of the result is loaded directly to RAM. If necessary, the result is also loaded to the S register to allow the high-order byte to be transferred to the RAM port.

4. *Neighbourhood operations.* In operations of this type the processor derives a result that is a function of its own value and those of its neighbours. The first step is usually to transfer a byte of data from RAM to Nout, and often simultaneously to the S register. Data are then transferred from each cell to its neighbours, each Nout to a set of Nin registers as shown in Fig. 5. In general, the input multiplexer then sequences around the inputs and, in each direction, the ALU computes a new result that is usually stored in the S register. Finally, the result is transferred byte by byte to RAM via the ALU.

These simple modes of operation form the basis of all circuit functions. The performance of the chip on a number of basic operations is shown in Table 2.

Table 2
Performance of the CLIP7 chip

Operation	*Execution time in μs*
Binary pointwise logical	1.6
8-bit pointwise add	1.6
8-bit pointwise multiply	14.5
Binary neighbour mask	1.6
8-bit neighbour maximum	8.8
8-bit neighbour median	40
Global propagation	1 (per pixel per bit)

3. PROPOSED SYSTEMS

One of the principal reasons for choosing the design outlined above was the flexibility it offered in terms of system design. Outlines of some of the systems that we propose to develop are presented in this section.

3.1 The CLIP7 system

In a field that seems beset with problems of nomenclature (Pattern Recognition? Image Processing? Image Analysis?) we have developed our own special difficulty, viz., the same name is applied to a custom circuit and to the first system in which it is to be used. Thus, in the present case, there exists a system design known simply as "CLIP7" [10]. This is the system intended to provide, in the first instance, a high-resolution equivalent to CLIP4. Its specification is given in Table 3, and its configuration is shown in Fig. 7. The important features of the system are as follows:

1. The processor array is a strip of 512 × 4 CLIP7 chips that scans across a 512 × 512 pixel data array.
2. Each pixel of data is provided with up to 256 bytes of local RAM storage. This means that each processor has up to 32 KB of directly accessible storage.
3. System data transfers are via a 32-bit bus to which all appropriate modules have access.

Table 3
Specification of the CLIP7 system

Data area	512 × 512 pixels
Data source	625-line TV camera
Operating speed	As CLIP4
Software	Upward compatible with IPC
Controller	ICL PERQ
Operating system	UNIX[a]
Control strategy	SIMD or MIMD
Local data storage	256 bytes per pixel (maximum)
Backup data store	1000 images
	100 ms access time per image
Display facilities	Software controlled
Additional facilities	Pixel addressing
	Bit counter

[a] *UNIX is a trademark of Bell Laboratories*

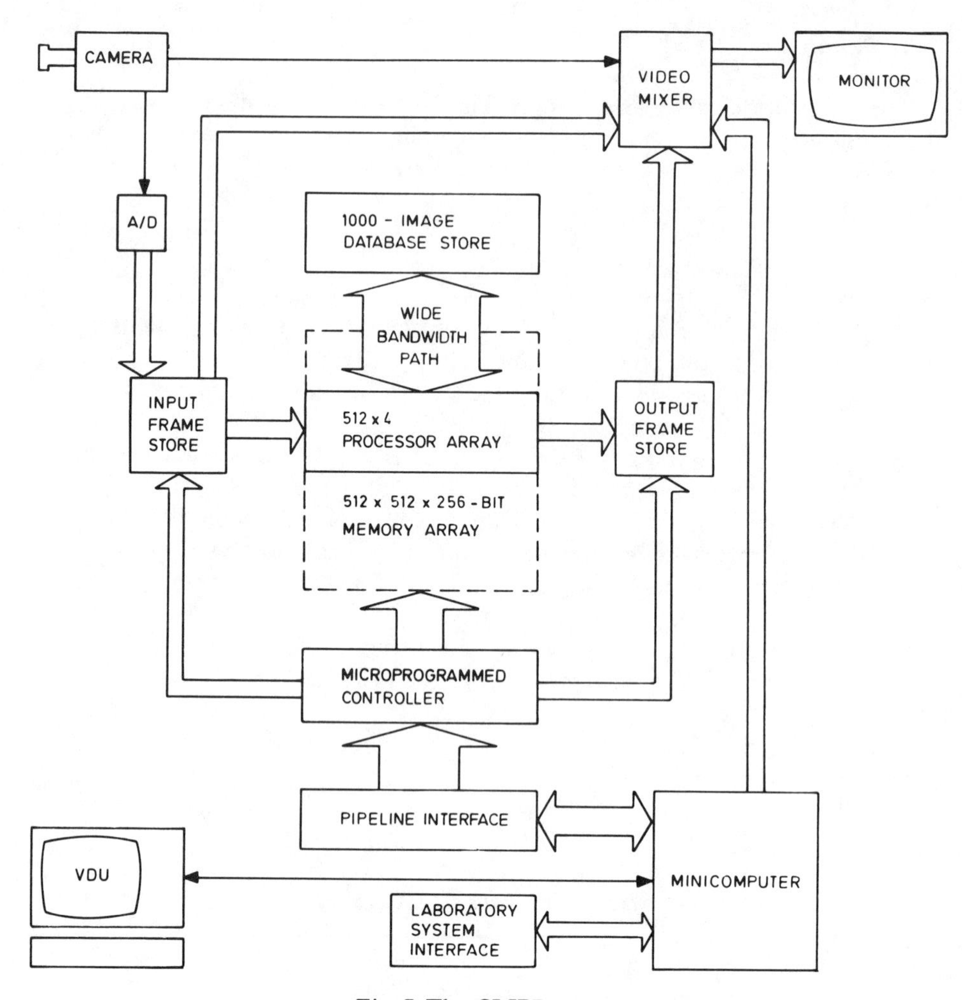

Fig. 7 The CLIP7 system

4. A database backup store is provided, based on a set of distributed winchester discs with cache memory interfaces, having a 20 MB per second data path to the array and able to store up to 1000 grey-level images.
5. System control is by means of a microcode sequencer interfaced to an ICL PERQ host computer. The microcode has two sections: a predefined set of standard routines and a user-programmable area for algorithm development.
6. The system operates in SIMD mode, except for the previously described local autonomy of processors.

Table 4
Performance of the CLIP7 system

Operation	*Execution time in ms*
Binary pointwise logical	0.2
8-bit pointwise add	0.2
Binary neighbour mask	0.2
8-bit neighbour maximum	1.2
8-bit 15 × 15 median	400
Frame store to array (8-bit)	14
Backup store to array (8-bit)	65

This system should facilitate two objectives: first, the extension of application studies to higher resolution images, and second, the investigation of algorithms that can exploit the local conditionality of the array. The performance of the system on some typical operations is detailed in Table 4.

3.2 An augmented array

The first step in extending the principle of local autonomy in an array beyond that intrinsically permitted by the CLIP7 chip is shown in Fig. 8. The intention is still to broadcast instructions to the whole array, but each processor now has complete local addressing facilities in its local data storage area. It would be appropriate to incorporate such an arrangement into a single-line scanned array to explore more fully the possibilities inherent in one processor having flexible access to all the data from a complete column of

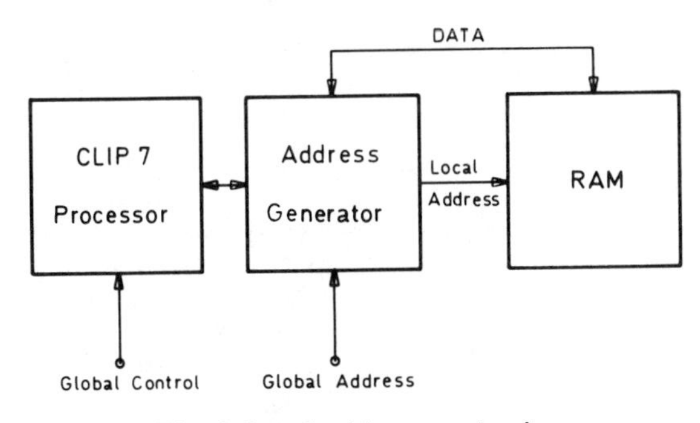

Fig. 8 Local address generation

pixels. In such a system it would also be feasible to incorporate a much richer lateral connectivity net to remove one of the serious limitations inherent in nearest-neighbour connected arrays.

3.3 A MIMD pyramid

The type of structure proposed here is not new in concept. Uhr *et al.* [3] and Tanimoto [4] have suggested ideas for pyramids of various types, Uhr in particular having considered the problem of MIMD control. However, as far as the author knows, only Tanimoto is involved in the construction of such a system, in which the control mode is SIMD.

Our feeling, reinforced by working with CLIP4, is that practical experience with a new architecture far outweighs prior analysis or simulation in obtaining an understanding of what it can achieve. Therefore, our intention is to construct a system on the scale of that shown in Fig. 9, having a small number of levels, each performing a separate program of operations. This type of structure has already been used conceptually in scene analysis studies, and Uhr has suggested that it might appropriately form the top section of a pyramid of the Tanimoto type.

Particular areas to be investigated here would be the partitioning of any problem/algorithm between levels, whether the pyramid structure imposes an absolute reduction in information toward the apex, and what degree of downward feedback might be appropriate, either in the array or between the level controllers.

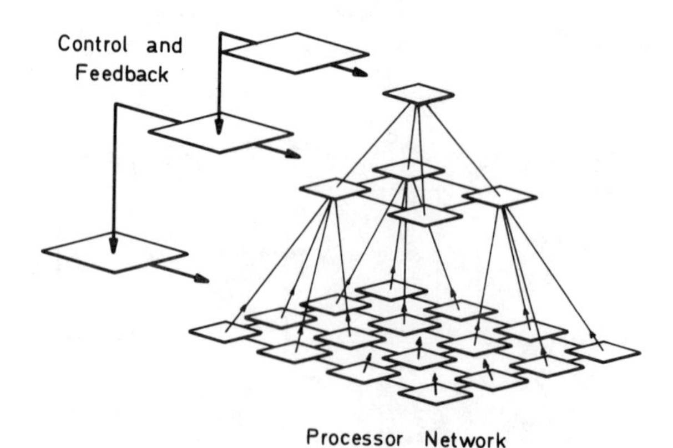

Fig. 9 Proposed MIMD pyramid structure

3.4 A cubic processor

Initial investigations of CLIP 2-D arrays took place using systems obviously too small to be of practical use but large enough to enable algorithms and control techniques to be developed. In a similar way, it is proposed to construct a small cubic lattice of CLIP7 chips, no larger than $8 \times 8 \times 8$ processors, having a flexible control distribution structure that would not limit the system to SIMD mode. Such a structure would emulate more closely the type of neural network that was one of the inspirations of many early processor network ideas and which, combined with some of the ideas involved in the field of expert systems, might lead to self-inferential or deductive networks.

It has to be emphasised here that such an investigation would be highly speculative, but that, hopefully, insights obtained during the use of some of the previously outlined structures would provide pointers to the way to proceed.

4. CONCLUSIONS

The program outlined in this chapter falls into two sections. In the first, we are attempting to produce quantitative improvements in the performance of a rather well-understood architecture, the SIMD array. In the second, we are attempting to move away from this structure in an evolutionary way to develop qualitatively different systems that are more appropriate to other ideas of control and data flow. During this process, we would hope to contribute to the development of such ideas. Both sections of the program are based on the use of the CLIP7 chip.

It is probably important to emphasise in a chapter of this speculative nature that an important aim of development programs of this sort should be practical. Thus, the development of CLIP4 was followed by a period of use of the system on various application studies and by the commercial availability of systems. In a similar way, the construction of the CLIP7 system described in Section 3.1 will be followed by its use in an extended application program. It remains to be seen if the ideas outlined here will prove useful in their turn.

REFERENCES

[1] Duff, M. J. B., and Levialdi, S. (eds.) (1981). *Languages and Architectures for Image Processing*. Academic Press, London.

[2] Duff, M. J. B. (ed.) (1983). *Computing Structures for Image Processing*. Academic Press, London.

[3] Uhr, L., Thompson, M., and Lackey, J. (1981). *IEEE Comp. Soc. Workshop on CAPAIDM*, 209–216.
[4] Tanimoto, S. L. (1983). *10th Internat. Symp, on Comp. Architecture*, Stockholm.
[5] Preston, K., Jr. (1982). Private communication.
[6] Rieger, C. (1981). *IEEE Comp. Soc. Workshop on CAPAIDM*, 119–124.
[7] Guzman, A. (1981). *IEEE Comp. Soc. Workshop on CAPAIDM*, 309–317.
[8] Dorey, A. P., Langstaff, D. P., and Jobling, D. T. (1983). *Technology Related Integrated Circuit Design Study* Southampton University, Microelectronics Centre Report.
[9] Fountain, T. J. (1981). *IEEE Comp. Soc. Workshop on CAPAIDM*, 25–30.
[10] Fountain, T. J. (1983). *Patt. Recog. Letters* **1**, 331–339.

Chapter Fourteen

Some Design Considerations on a Fast, Reliable, and Low-Power Multiprocessor System for Image Processing on Board Scientific Satellites

Frans A. Gerritsen, Anton Monkel, and Hans F. A. Roefs

1. INTRODUCTION

Many of the scientific satellites that are in use today or foreseen for the future involve imaging sensors. For a number of reasons, it is often advantageous to apply image-data compression (source-encoding) techniques on board such satellites. However, the requirement that data compression should keep pace with data acquisition results in processing-power requirements that cannot be met by today's general-purpose on-board computers. Therefore, in current scientific imaging satellites, image-data compression is either absent or handled by dedicated, nonprogrammable hardware. The cost of adaptation of this existing, dedicated image-compression hardware to new sensors and/or to new compression techniques often turns out to be comparable to the cost of designing and building a new, dedicated image-compression system.

The main objective of the CADISS project, on which this chapter reports, is to design and build a so-called Elegant Breadboard of an adaptable and programmable image-compression and decompression system for use on board scientific satellites (CADISS is an acronym for Compression And Decompression of Imaging Sensor Signals). The project should demonstrate the feasibility of a system that can be programmed with a number of well-performing image compression/decompression algorithms to process image data at relatively high throughput rates with relatively low power

INTEGRATED TECHNOLOGY FOR PARALLEL IMAGE PROCESSING
ISBN 0-12-444820-8

consumption. The system should be radiation tolerant and should fit in a small volume.

Some of these design objectives may be conflicting: A programmable system is usually slower than a system that uses dedicated hardware to perform its functions; high-speed processing devices usually consume a considerable amount of power; the high packing density that is required to achieve a small volume may not be possible with the electronic devices available. It is the purpose of this chapter to show why the current CADISS multiprocessor design is an attractive compromise.

An outline of the CADISS system has been given earlier by Roefs [1], and Huisman [2] has reported on the image-data compression algorithms that will be used in the system.

2. ADAPTABILITY: DESIGN GUIDELINES

We tried to enhance the adaptability of the CADISS multiprocessor image-data compression system to the requirements of new satellites by adopting in our design the following guidelines:

1. A range of types of camera instruments that are currently in use should be hosted without (or with slight) modifications. Adaptation of CADISS to new camera instruments should involve modifications of the instrument interface only (see Section 2.1).
2. Several operating configurations for on-line and off-line compression and decompression should be available (see Section 2.2).
3. Both fixed-error and fixed-compression–factor modes of the various image-data compression algorithms should be available (see Section 2.3).
4. Within certain acceptable limits, it should be possible to adapt the CADISS system to higher pixel rates and/or to more computation-intensive algorithms by incorporating more processor modules (see Section 2.4).

In Section 3 of this chapter, some of the technological requirements are discussed. Two ways of partitioning the image-compression workload over a number of processors are compared in Section 4. In Section 5, a description is given of the CADISS multiprocessor architecture. The projected characteristics of the CADISS system are summarized in Section 6.

2.1 Imaging sensors on board satellites

The range of types of on-board camera instruments that are in use today can be classified according to the dichotomy snapshot/scanning instruments or according to the dichotomy monospectral/multispectral instruments. In most cases, both the snapshot and the scanning types of on-board camera instruments deliver their image data as a line-by-line bit-serial or byte-serial stream.

Four major types of multispectral camera instruments can be distinguished, according to the ways in which they interleave (mix) the data from the various spectral channels (or bands) into a single stream: the pixel- , line- , swath- , and image-interleaved types of multispectral instruments.

When using pixel interleaving, the instrument alternates first in the channel (spectral) direction, then in the image-line (row) direction, and finally in the image-column direction.

Line interleaving is achieved by alternating first in the image-line direction, then in the channel (spectral) direction, and finally in the image-column direction.

Swath interleaving is sometimes used when a sensor with a linear (N-element) receptor-array is involved (placed in the column direction, scanning in the line/row direction). In that case, the instrument first delivers a swath of N lines of the first channel (spectral band), then the same N lines of the next channel, and so on, until all channels of the current N lines have been delivered. This sequence is repeated for all N-line swaths of the image.

When using image interleaving, the instrument sequentially delivers "complete" images of the various channels.

To be adaptable to such a range of camera instruments, the instrument interface must be complex. The interfacing is further complicated by the requirement that for multispectral sensors it should be possible to use multispectral compression algorithms. Such algorithms (typically) operate on $N \times N \times K$-pixel blocks, with N columns and N rows in the spatial directions and K bands in the spectral direction.

For the CADISS system, we decided to use two N-line, K-band, 1 byte/pixel input data buffers ($N=8$ or $N=16$; $2\,N\,K \times$ line length $\leq 131,072$) which are alternatingly filled by the camera instrument and emptied by the processors in a ping-pong fashion. The CADISS instrument interface is capable of handling the various serial-to-parallel conversions necessary for the instrument handling in the line- and image-interleaving approaches. Monospectral instrument data can be handled as a special case of multispectral data. By treating each monospectral instrument as a spectral channel, several such instruments can be handled simultaneously. The handling of pixel- or swath-interleaving would require an adaptation of the CADISS instrument interface.

2.2 Operating configurations for compression and decompression

To enable the CADISS system to be used for a variety of satellite missions, a number of operating configurations for image-data compression and decompression should be available.

For a large number of satellites, the on-line, direct-transmission way of compression will be sufficient. In this operating configuration, the image data are compressed and transmitted immediately upon acquisition. Of course, this requires

(a) sufficient power at the time of acquisition to allow the compression processors to function,

(b) sufficient power to allow the communication down-link to function,

(c) communication between satellite and receiving ground station (clear line-of-sight from satellite to ground station is not always available when nongeostationary satellites are used),

(d) a receiving ground station that is sophisticated enough to decompress the received image data.

If these requirements cannot be satisfied under all circumstances, then the use of other operating configurations may be advantageous.

In the on-line-compression, deferred-transmission operating configuration, image data are compressed immediately upon acquisition, but the compressed data are stored in on-board mass memory, waiting for sufficient power (or for visibility of the ground station) for the down-link to be established. Compression before storage in mass memory allows acquisition of more image data (for a certain size of mass memory).

In the off-line-compression operating configuration, image data are directly stored in mass memory, waiting for later compression and transmission. This configuration is useful when power is not sufficient to allow the compression processors to function.

In the on-line-compression, off-line-decompression configuration, image data are compressed immediately and stored in on-board mass memory. When the down-link is established, data are decompressed and transmitted to the receiving ground station. This configuration is useful if one wishes to have the advantage of the acquisition of larger amounts of image data during the periods that the down-link is not available, even if the ground station is not sophisticated enough to decompress the image data.

The CADISS system will be capable of operating in all of the on-line configurations described above. Operation in the off-line compression configuration would be possible after an adaptation of the instrument interface.

2.3 Modes of algorithm operation: Fixed error and fixed compression factor

For some applications, the end-users of the image data acquired by satellite require that image data should be compressed and decompressed without reconstruction error. Entropy coding makes it possible to do so.

For some other applications, it is required that a certain fixed (RMS) reconstruction error can be guaranteed for the decompressed data. This calls for the *fixed-error mode* of operation of the compression algorithms.

On the other hand, the nature of some satellite missions (especially deep-space probes) requires that it should be possible to transmit each image with a certain fixed (maximum) number of bits. This calls for the *fixed-compression-factor mode* of the compression algorithms.

Sometimes these modes of operation are alternated: The bulk of imagery is transmitted with a fixed compression factor; from this imagery, one chooses areas of interest to be rescanned and transmitted in the fixed-error mode or in entropy-coding mode.

Currently, three image-data compression algorithms are available for the CADISS system. The Rice algorithm [3] provides for entropy coding, whereas a (proprietary) ESTEC algorithm (by Chaturvedi) and (modified) Melzer algorithm [4] both provide for compression with a fixed (RMS) reconstruction error at block level. All three algorithms can operate in a mode in which a fixed compression factor is guaranteed at the block level.

Each of the algorithms is preceded by some decorrelating transform: Hadamard, Walsh/Hadamard, or discrete cosine transform (DCT). More details have been given by Huisman [2].

2.4 Multiprocessor design guideline

It should be possible to adapt the CADISS system to higher pixel rates and/or to more computation-intensive algorithms by incorporating more processor modules.

To enhance the compatibility of the CADISS multiprocessor architecture with other multiprocessor on-board computer architectures in development for ESA–ESTEC (see, for example, Gaillat [5]), we decided to interconnect the various modules of the CADISS system by means of a ring bus. (A discussion of a number of computer architectures that have been proposed or which are in use for image-processing applications can be found in Gerritsen [6].)

3. TECHNOLOGICAL REQUIREMENTS

3.1 Power consumption

A general requirement for on-board electronics is that the power consumption should be as low as feasible: Both the supply of electric power to on-board electronics and the removal of dissipated heat are costly.

There are several ways to limit the required power supply of an compression/decompression system such as CADISS.

One can, for example, try to simplify further the algorithms, while retaining their performance, so that fewer processors are needed, or so that slower clock-cycle times can be used. In the CADISS project, extensive efforts have been made to assure that the powerful algorithms can be executed in a simple way.

The (dynamic) power consumption is also a function of the required throughput rate and of the technology used for processors, memories, and input/output hardware. The CMOS technology offers an atttractive combination of low power consumption and relatively high speed (see also Sections 5.2.4.2 and 5.2.4.4).

3.2 Operational temperature range

Although CADISS [being an Elegant Breadboard (EB)] needs not meet space-environmental requirements, the potential for converting the EB to a flight unit should be present. Therefore, all electrical components will be capable of operating in the industrial temperature range (−40–+85°C).

3.3 Radiation tolerance

To facilitate conversion of the CADISS EB to a flight unit, we decided to avoid the use of integrated-circuit devices of technologies that do not have a potential for radiation hardening (see also Section 5.2.4.4).

4. PARTITIONING OF COMPUTATIONAL WORKLOAD AND CONTROL

When designing a multiprocessor computer architecture, one has to consider the ways in which the computational workload can be partitioned over the processors. One also has to consider the partitioning of control over the processors, the processor-interconnection scheme following from the chosen partitioning of the workload, and the transfer rates required for the processor–interconnection structure.

The relationships and dependencies between the subtasks that are distinguishable in the workload impose limitations on the freedom the designer has in partitioning the workload over a number of processors. In the image-data compression algorithms that have been considered for CADISS, two ways of partitioning the workload into subtasks with little interdependency can be employed: a functional partitioning, which follows a functional division of the workload into subtasks, and a data partitioning, which divides the workload into subtasks in the same way as the data are divided into data-blocks.

These two types of partitioning will be discussed and compared in the following subsections. (Although the CADISS architecture does not prohibit the use of the functional-partitioning approach in any way, it was decided to use the data-partitioning approach in the implementation of the various algorithms for CADISS.)

4.1 The functional partitioning approach

In the functional partitioning of the image-data compression workload, the subtasks may, for example, be chosen to be (see Fig. 1):

1. decorrelating transformation (e.g., 2-D–DCT) of an $N \times N$-pixel block;
2. determination of bit allocations for each of the transform coefficients;
3. compression of the bit-allocation table;
4. quantization of the transform coefficients and formatting of the quantized coefficients and the bit-allocation table for transmission to the ground.

Apart from the obvious precedence dependencies, these subtasks are independent of each other. A continuous flow of $N \times N$-pixel blocks can be processed in a pipelined fashion, for example, by four processors, with each of the processors assigned to a separate subtask. Pipeline interconnections are

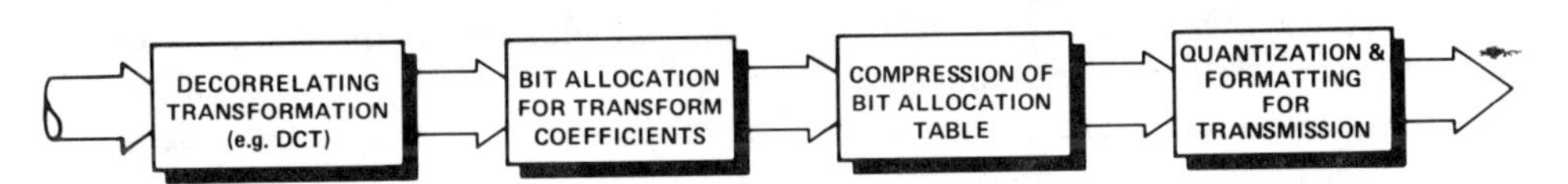

Fig. 1 Functional diagram of a typical image-data compression algorithm as considered for the CADISS system.

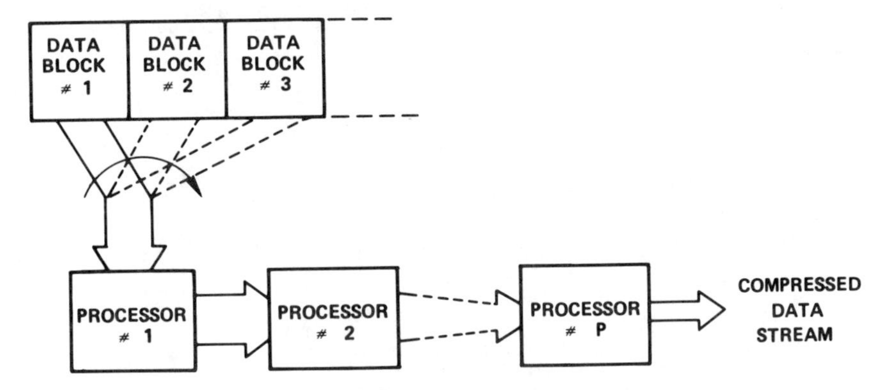

Fig. 2 The *functional partitioning* approach to the division of the image-data compression workload over a number of processors.

then needed between processors to transfer intermediate results (see Fig. 2). Conceivably, one could use several of such quadruplets of processors if the throughput of a single quadruplet turns out to be insufficient.

4.2 The data-partitioning approach

In the data-partitioning approach, each processor executes the complete data-compression algorithm for an $N \times N$-pixel block. Separate processors handle separate blocks of data (see Fig. 3). Conceivably, instead of using a set of isolated processors, one could use a set of pairs (or triplets or quadruplets) of processors.

4.3 Comparison of the functional-partitioning and the data-partitioning approaches

When one compares the functional-partitioning approach with the data-partitioning approach, one notes the following advantages and drawbacks of each of the approaches:

1. In the functional-partitioning approach, the data rate of the inter-processor bus ought to be relatively high. In addition to the transfers of incoming data to the first processor of the pipeline and of the resulting compressed data from the last processor to the output interface, intermediate results have to be carried between adjoining processors.

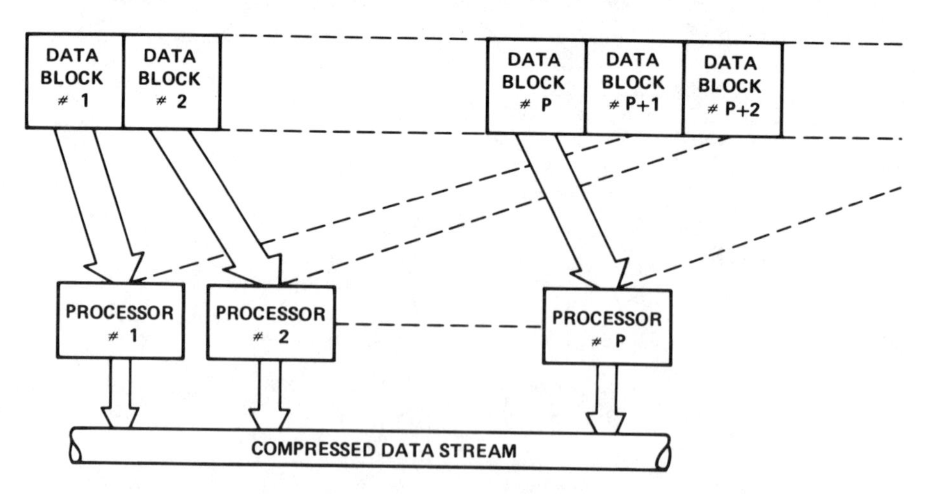

Fig. 3 The *data partitioning* approach to the division of the image-data compression workload over a number of processors.

2. The number of processors expected to be required for a certain data rate is higher in the functional-partitioning approach, because the transfer of intermediate results may cost considerable time.

3. In the data-partitioning approach, the program memory of each processor should be larger, because each processor must be able to perform a complete data-compression algorithm.

4. In the data-partitioning approach, there are more opportunities for incorporating redundancy (and graceful degradation) into the system, because all processors are necessarily identical. (Of course, in the functional-partitioning approach one also could choose the processors to be identical.)

5. For the same reason, the implementation of the data-partitioning approach is easier and more simple. Also, the functional-partitioning approach is harder to implement, because each functional subtask should be executed in the same time.

6. The data-partitioning approach allows easier adaptation of the multiprocessor system to higher (and lower) data rates, because more (or fewer) processors can be included in a straightforward way. In the functional-partitioning approach, one would have to resort to (additional) data-partitioning methods to allow for higher data rates. In both cases, lower data rates can be handled by using a slower processor-clock frequency.

7. With regard to the fixed-compression-factor mode of the data-compression algorithms, one should note that the data-partitioning approach does not lend itself to easy implementation of compression algorithms which require that locally under-average performance of the compression should be compensated for by over-average performance on the $N \times N$-pixel block that adjoins in the row direction. (This would impose a precedence constraint on the compression of adjoining blocks.) On the other hand, such adaptive fixed-compression-factor algorithms can be readily implemented if one allows the compensation to be travelling in the column direction (provided that blocks that adjoin in the column direction are handled by the same processor).

Because of the considerations mentioned here, we decided to use the data-partitioning approach for the CADISS multiprocessor system.

5. DESCRIPTION OF THE CADISS MULTIPROCESSOR ARCHITECTURE

5.1 Description at the system level

The processors, interfaces, and memory units of the CADISS multiprocessor are interconnected by means of a ring bus (see Fig. 4). Both image data and control information are passed through this ring bus.

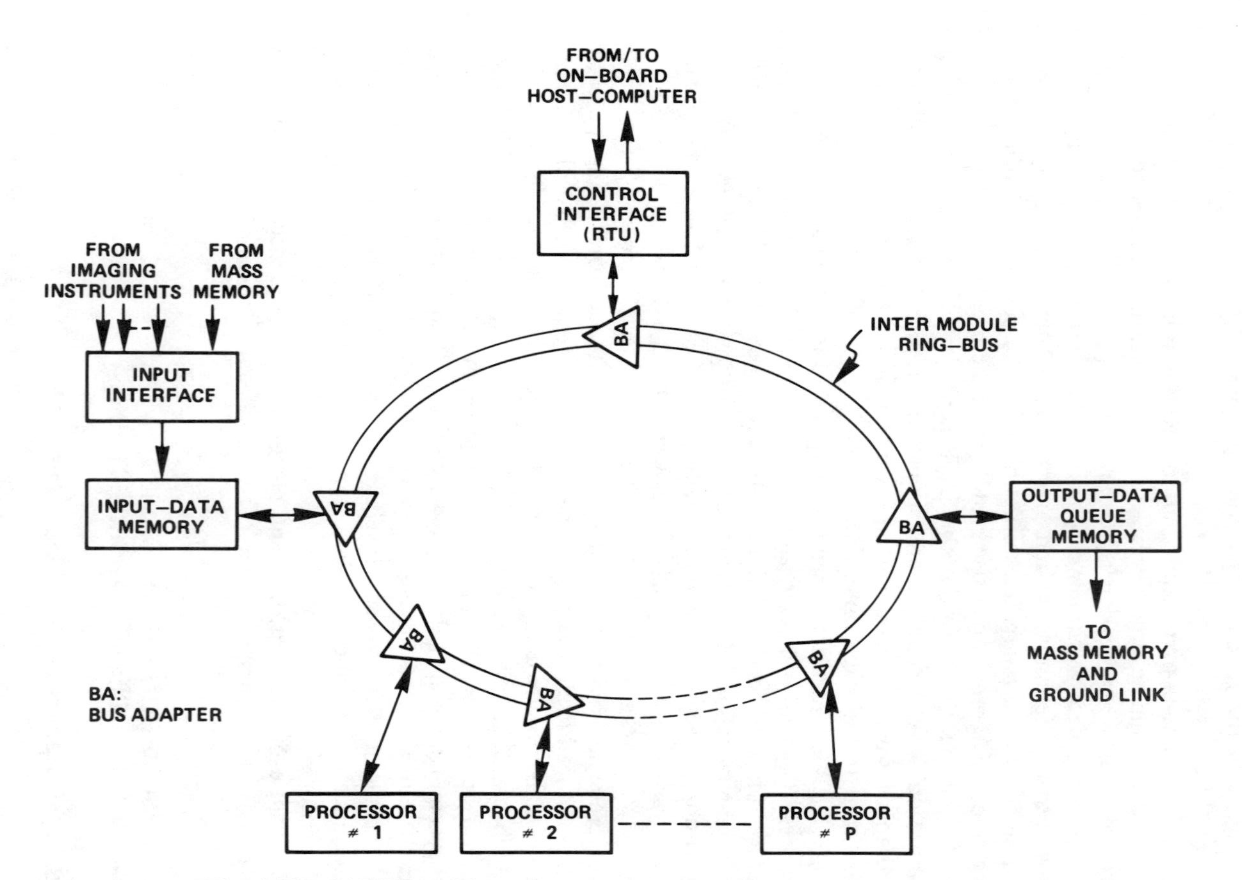

Fig. 4 The CADISS multiprocessor architecture with ring-bus interconnections.

The microcode of the selected algorithm, together with the various coefficients, parameters, tables, and format specifications must be loaded by the host computer (i.e., the on-board data-handling system) through the remote terminal unit (RTU) control interface. Although the set of available algorithms will normally be stored in on-board mass memory, this configuration (in principle) allows programs to be sent to the satellite during flight (e.g., to correct possible programming errors or to incorporate new algorithms).

The ring bus can be thought of as a bucket-brigade loop. Each bucket holds 16 bits of data and 8 bits of status/address information. A bucket can be empty (available for use) or filled (on its way to its destination). At the start of a ring-bus clock cycle, each bucket is passed to the bus-adapter of the next module on the ring bus (in the anticlockwise direction in Fig. 4).

When a module needs to transmit a datum to another module, it waits for an empty bucket to pass its bus-adapter, places the address of the destination module and a 16-bit datum in the bucket, and marks the bucket *filled*.

All bus-adapters continually monitor the status/address part of each passing bucket, comparing the bucket's destination address with their own. If these addresses correspond, the incoming datum is copied to a memory internal to the module, and the bucket is marked *empty*.

It is also possible to "broadcast" messages. (For the sake of brevity, the broadcasting procedure is not described here.)

The data transfers to and from the system can be controlled by external clock signals generated by the data source (say, an instrument) and data sink (say, data storage) or by internal clock signals generated by the system. The data transfers are enabled and disabled by handshake procedures.

On the instrument (sensor) side, both a monoinstrument (multichannel) and a multiinstrument option are available. Incoming data are divided into 8×8- or 16×16-pixel blocks and passed on to the respective processors.

On the data-output side, the data are formatted in blocks of frames. Each block of frames contains the compression results of 8 or 16 image lines. Each output frame contains the compression result of an 8×8- or 16×16-pixel block.

5.2 Description of the various modules

5.2.1 Sensor instrument-to-system interface

The CADISS system will be capable of handling four different instruments. Two options will be available: multi-instrument and (multi-channel) mono-instrument.

In the multi-instrument option, the selection of an instrument is initiated by a port enable. The instrument-selection sequence is dictated through a format-control command from the RTU control interface. The actual port commutation is controlled by the image-sync signal (in the case of image interleaving) or by the line-sync signal (in the case of line interleaving).

In the mono-instrument option, only data transfers to a single port are possible. When the instrument is of the multi-channel type, the data can be presented in the image-interleaved sequence or in the line-interleaved sequence. However, commutation of the images (channels) must be done within the instrument. The number of interleaved lines (maximum of eight) is dictated by the RTU control interface through a format-control command. The instrument interface will arrange a sequence of lines into one concatenated line, which will then be treated by the system as one image line.

The total number of pixels of a line (or of a concatenated line) should not exceed 4096 and must be a multiple of 512. The total number of lines of an image must be a multiple of 8 (if 8×8-pixel blocks are used in compression) or 16 (if 16×16-pixel blocks are used). In the image-interleaved mode, the number of lines per image can vary from instrument to instrument (channel to channel). In the line-interleaved mode, however, the images must contain the same number of lines.

5.2.2 Memory modules

As is shown in Fig. 4, memories are placed between input interface and ring bus, and between ring bus and mass memory/ground link.

The input-data memory contains 64 RAM-chips of 2048×8 bits each. It is divided into two parts of equal size (64 KB each), which are alternately filled by the sensor instrument (s) and emptied to the processor modules through the ring bus. Each section can contain the image data of 16 lines of 4096 pixels each (8 bits per pixel). The sections change roles when 16 lines have been received from the instruments (or 8 lines when 8×8-pixel blocks are used in the compression).

Data are transported to the processor modules in 8×8-pixel or 16×16-pixel blocks. Each block is sent only to the processor module that will try to compress it. To facilitate these transfers, the data from the sensor instruments are stored in such a way that a block can be read with linearly incrementing addresses.

The output-data–queue memory contains two RAM-chips of 2048×8 bits each and is configured as a *first-in, first-out* memory. This output memory can contain the compressed results of 64 8×8-pixel blocks or 16 16×16-pixel blocks.

5.2.3 RTU Control interface

The on-board data-handling system (OBDH) is used as the host computer of the CADISS system. The host computer exchanges command and status information with the CADISS system through the remote terminal unit (RTU) control interface. The commands that can be issued by the host computer include

(a) *reset* to down-load the microcode of the processor modules in the ring bus broadcast mode,

(b) *initialize* (also broadcasted), to specify coefficients and parameters of the compression algorithm,

(c) *configuration* to specify the operating configuration (also specifies block size, compression/decompression, and availability of processors),

(d) *start* to start the compression process,

(e) *stop* to stop the compression process,

(f) *test* to test the correct functioning of the system (a special data block is generated and processed by all processors, the result of the process is verified, and a success/failure report is returned to the host computer).

5.2.4 Technological aspects of the processor module

In the current design of the CADISS processor modules, special attention has been given to programmability, high speed, low power consumption, small volume, potential for radiation hardening, and a wide temperature range, all within the limits of the components available in mid-1983.

5.2.4.1 Programmability and high speed

The requirement of programmability rules out the use of dedicated hardware. Microprocessors, microcomputers, and signal processors all have the potential of programmability. Each CADISS processor module uses a bit-sliced microprocessor and a multiplier–accumulator with writable memory (RAM) for storage of programs and microcode. A program reload is needed after a complete power-down of the system.

The combination of a bit-sliced processor with a separate multiplier–accumulator (and separate memories and interconnecting internal buses) was favored over the alternative of a digital signal processing (DSP) chip, mainly because of the limited memory and I/O-facilities of the current DSP-chips. Also, the current DSP-chips are not available in technologies that have potential for radiation hardening. A disadvantage of the current microprogrammable design is the cumbersome translation from algorithm to microcode during the development of the application software.

5.2.4.2 Low power consumption and high speed

As is well known, the power consumption of integrated circuits depends on the technology used in their fabrication. Relative to the power consumption of ordinary TTL-technology, the power consumptions of the most common technologies are approximately: ECL (10), Schottky TTL (5), TTL (1), low-power Schottky TTL (0.2), NMOS (0.1), and CMOS. CMOS has a very low (order of microwatts) quiescent power consumption. The dynamic power consumption of CMOS devices increases with the operating frequency.

To obtain low power consumption, one should try to use as many pure CMOS components as possible, as long as a certain operating frequency is not exceeded. Memory devices are available in CMOS/NMOS and in pure CMOS technology. Focussing on 1K × 4 and 2K × 8 static RAM devices, one obtains the following mid-1983 picture of the market:

CMOS/NMOS: access time 55–100 ns, power consumption 200–300 mW;

CMOS: access time 200–300 ns, power consumption 20–30 mW/MHz; Matra/Harris offers a 2K × 8 RAM with an access time of 100 ns and power consumption of 20–30 mW/MHz (other manufacturers have announced similar devices).

With regard to *glue-logic* devices, a growing (but not complete) family of CMOS logic circuits is available with speeds comparable to their low-power Schottky TTL pin-compatible forefathers (gate delay approximately 10–20 ns) but with a drastically reduced power consumption. The older 4000 family of CMOS logic has a gate delay of about 100 ns and is, therefore, not practical in fast systems.

5.2.4.3 Small volume

The cabinet that has been proposed for CADISS allows about 16 printed circuit boards, with a usable area of 7.5 cm × 16.5 cm each. Each processor module must be assembled on one board. This is possible only when one uses leadless chip carrier packages (LCCs) for most circuits. Only a few types of memory, glue-logic, and processor devices are currently available in LCC: Matra/Harris offers a 2K × 8 RAM with an access time of 100 ns and power M10900 in LCC, and RCA offers the EPIC processor in LCC. Processor-oriented logic such as drivers, transceivers, registers, latches, and decoders are available in LCC from Mitel. Other circuits are not currently available in LCC.

5.2.4.4 Potential for radiation hardening

As is well-known from literature (see, e.g., [7]), large-scale and very-large-scale integrated circuits (LSI and VLSI) in NMOS technology are rather susceptible to soft and hard cell and gate errors, caused by heavily ionized cosmic-ray nuclei. Devices of technologies such as CMOS–SOS (silicon-on-sapphire) can, however, be hardened to allow very high total dose levels.

5.2.4.5 Temperature range

The extended temperature range of −55–+125°C, which is usually required for space-qualifiable components, poses a problem in the following areas:

Modern, fast, CMOS glue logic is currently available only in plastic DIL packages with the industrial temperature range of −40–+80°C. However, most manufacturers expect that the CMOS logic family will be available in ceramic packages with the military temperature range.

Leadless chip carriers (LCCs) cannot be mounted on normal glass epoxy boards because of the difference in thermal behaviour of the epoxy board and the ceramic LCCs.

As a solution to the latter problem, two options come into focus:

The LCCs are mounted on ceramic multi-layer substrates which, in turn, are mounted on the epoxy printed-circuit boards.

The LCCs are directly soldered onto a special type of printed-circuit-board material, with a thermal behaviour close to that of the LCC ceramic.

5.2.4.6 Component selection

The above considerations led us to select the following components:

GP001 RCA EPIC 8-bit wide processor slices and GP502 RCA EPIC sequencer (both CMOS–SOS LCC);

ADSP-1010 Analog Devices 16×16-bit multiplier–accumulator in CMOS flat-pack;

HM-6516-B Matra/Harris 2K×8 static memory (100 ns access) in CMOS LCC;

54SCxxx-series Mitel glue logic (20–30 ns gate delay) in ISO–CMOS LCC;

74HCxx-series Motorola and National Semiconductor glue logic (10–15 ns gate delay) in high-speed CMOS, plastic DIL package −40–+85°C.

5.2.5 Description of the processor module

In Fig. 5, a simplified block diagram is given of one processor module of the CADISS multiprocessor system.

Through its bus-adapter, each processor module can communicate with all other modules connected to the ring bus (see also Section 5.1). Both module-to-module transfers and broadcast transfers are possible. The bus-adapter contains the electronics to decode the status/address contents of the buckets passed through the ring bus. It is also responsible for clearing to *empty* the status of buckets addressed uniquely to the processor and for encoding status information for buckets to be sent by the processor. Still, data transfers are performed in a programmed I/O fashion: During actual data transfers the bus-adapter is controlled by the microcode of the processor.

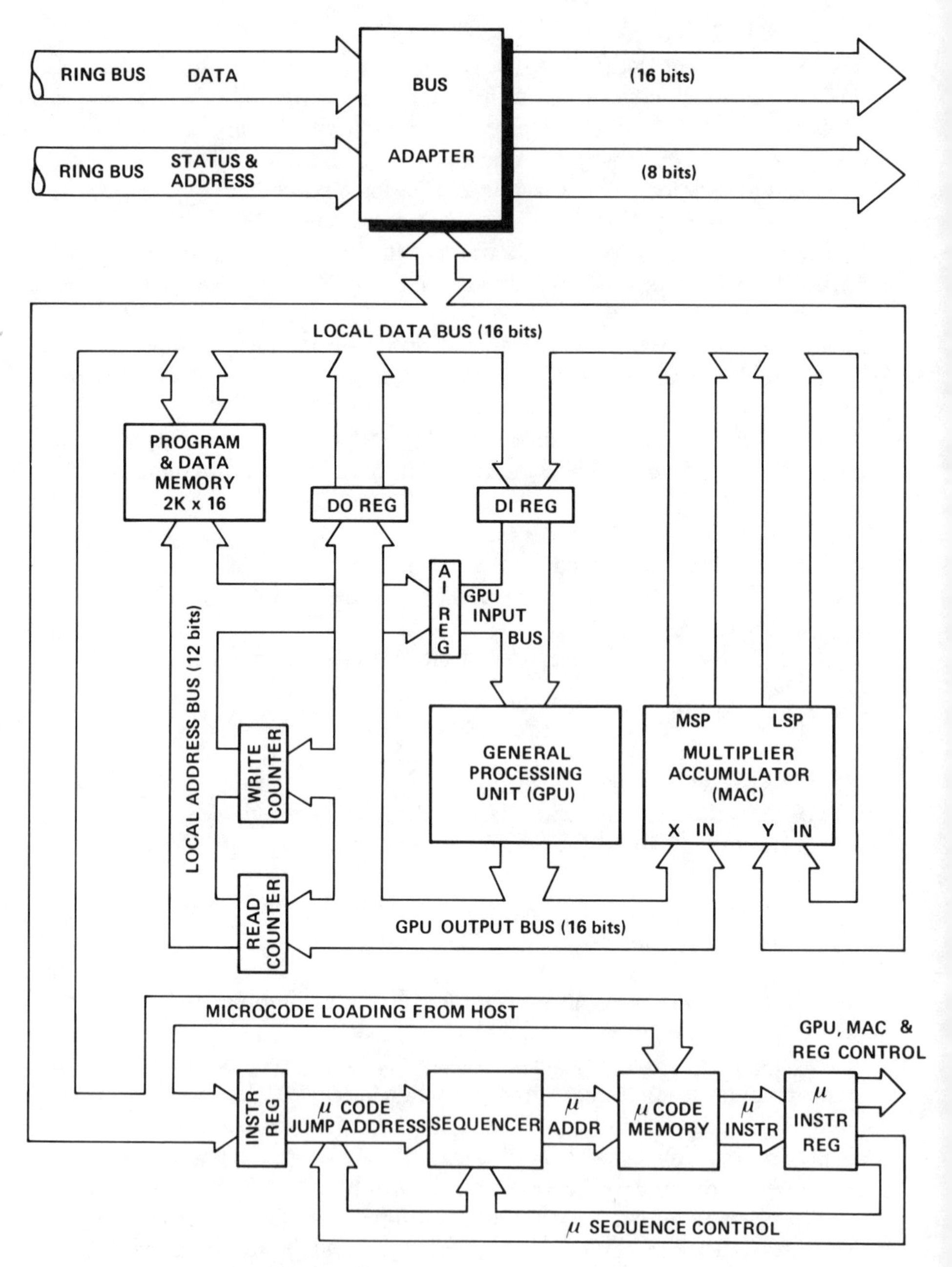

Fig. 5 Simplified block diagram of one processor module of the CADISS multiprocessor architecture.

The processor itself consists of an arithmetic section, a memory section, and a control section.

The arithmetic section contains a 16-bit general processing unit (GPU) consisting of two 8-bit GP001 RCA EPIC slices and an ADSP-1010 Analog Devices 16 × 16-bit multiplier–accumulator (MAC).

The GPU can perform arithmetic (addition and subtraction), logic, and shift operations on the data in its 16 internal registers and on data supplied through the GPU-input bus. The resulting data can be stored in the internal registers or can be passed to other units through the GPU-output bus.

The MAC can perform multiplications or sums of multiplications. The X-input of the MAC is connected to the GPU-output bus, so that the MAC can operate on GPU results. The Y-input of the MAC is connected to the local data bus, which allows memory data to be fed to the MAC. Memory data can also be fed to the X-input if the GPU is programmed to pass memory data to the MAC's X-input. The 35-bit result of the MAC can be passed over the local data bus in 16-bit parts.

A 2K×16 RAM is used as data/program memory. Incoming data (such as the 8×8-pixel or 16×16-pixel blocks of sensor data), coefficients, parameters, and look-up tables are stored here, together with the intermediate and final results of the compression algorithms. The macro program is also stored in this memory.

The memory data can be sent to the GPU-input bus (through the local data bus and the DI-register, which couples the local data bus to the GPU-input bus), to the MAC Y-input, to the macro instruction register, and to the ring bus (through the bus-adapter). Data can be sent to the memory from the GPU-output bus (through the DO-register, which couples the GPU-output bus to the local data bus), from the MAC output registers, and from the ring bus (through the bus-adapter).

During each microinstruction cycle, the memory may be accessed twice: for a *write* action and for a *read* action. The write and read addresses are supplied to the memory by the write-address and read-address counters. These counters can be loaded from the GPU-output bus and can be incremented, decremented, or kept unchanged under microprogram control.

The control section contains the macro instruction register, the GP502 RCA EPIC sequencer, a 2K×64 microcode memory, and a 64-bit micro-instruction pipeline register. The major part of the 64 bits of each micro-instruction are used to control the functions and clock-enables of the various components of the processor module. A small part of each 64-bit micro-instruction is fed to the microprogram sequencer to control the flow of the microprogram, i.e., to control the computation of the address of the next microinstruction to be executed. Certain next-address computations are subject to microprogram-selected condition signals (or flags; not shown in Fig. 5). The sequencer can perform such functions as stepping, looping, branching, and microsubroutine entry and exit.

Microprograms are down-loaded by the RTU control interface (see also Fig. 4) through the ring bus and the local data bus into the writable microcode memory of the control section of each processor module.

The data transfers to/from the processor module through the ring bus are controlled by the processor's control section (microprogrammed I/O). From the viewpoint of ring-bus data transfers, each processor can be in one of three different modes: the active mode (in which the processor wants to send or is sending data to one or more other ring-bus modules), the passive mode (in which the processor expects data from another ring-bus module), or the idle/executing mode (in which the processor is busy processing previously received data). In the active and passive modes, the power consumption is small, because clock cycles are fed to the processor only when needed.

6. PROJECTED CHARACTERISTICS OF THE CADISS SYSTEM

6.1 Operating characteristics

The CADISS system will perform compression or decompression of image data in accordance with a selection of predefined algorithms, parameters, and formats. The various options can be summarized as follows:

Algorithm modes	Fixed-error or fixed-compression-factor compression and decompression; self-test mode
Available algorithms	DCT + Chaturvedi; Walsh–Hadamard + Chaturvedi; DCT + Melzer; Walsh–Hadamard + Melzer; Rice (8×8 only)
Processing block sizes	8×8- or 16×16-pixel blocks
Input interleaving	Multiinstrument, image- or line-interleaved; single instrument (multichannel) image- or line-interleaved.
Clocks	Internal or external

The system design is such that compression of 8×8-pixel blocks in the fixed-compression-factor mode of the DCT + Chaturvedi algorithm can be performed with a maximum input rate of 4.8 Mbit/s (external clock) or 4 Mbit/s (internal clock).

The (tentative) microinstruction clock-cycle time will be 250 ns. The (tentative) execution time for an 8×8-pixel 2-D-DCT will be 187 μs (for a single processor board; this includes the microinstructions needed to keep round-off errors to a minimum but excludes the time (21 μs) needed to transfer the 8×8-pixel operand data block from the input-data memory to the processor board's internal data memory).

6.2 Physical characteristics

The Elegant Breadboard of the CADISS system will have the following (tentative) physical characteristics:

Dynamic power consumption	Will not exceed 12 W (total)
Operating temperature range	-15–$+50$°C
Total mass	Will not exceed 8 kg
Total volume	Will not exceed 8 dm^3
IC-technology	Pure CMOS or CMOS–SOS

7. CONCLUSIONS

Some of the considerations that have led to the current design of the CADISS multiprocessor system for image-data compression and decompression on board scientific satellites have been discussed. In the design, much attention has been given to the enhancement of the adaptability of the system to the requirements of a range of scientific satellites.

The feasibility of an adaptable and programmable on-board compression and decompression system having been shown, we are now awaiting approval for starting the construction of an Elegant Breadboard of the system.

A large number of the features of the CADISS system are also useful in ground-based applications. As a separate effort, the construction of a simplified version of the system for ground-based applications is being considered.

ACKNOWLEDGEMENTS

The research and development on which this chapter reports were carried out by the National Aerospace Laboratory NLR under European Space Agency ESTEC contract 5047/82/nl/hp and Netherlands Agency for Aerospace Programs (NIVR) contract 1987.

The opinions and views expressed by the authors of this chapter are solely their own and, therefore, do not necessarily represent those of the NLR, ESA–ESTEC, or NIVR.

The authors wish to thank D. C. Chaturvedi (ESA–ESTEC) and J. C. A. van der Lubbe, W. C. Huisman, and J. G. Koning (all of the NLR) for their contributions, suggestions, and stimulating discussions.

REFERENCES

[1] Roefs, H. F. A. (1983). CADISS: an image (de)compression system for deep space application. *Proc. Fourth Symposium on Information Theory in the Benelux, Catholic University of Leuven*, Leuven, Belgium, 26–27 May, pp. 121–127.
[2] Huisman, W. C. (1983). Three image compression algorithms for CADISS. Presented at the *Fourth Symposium on Information Theory in the Benelux, Catholic University of Leuven*, Leuven, Belgium, 26–27 May (available from the authors).
[3] Rice, R. F. (1979). Practical universal noiseless coding. In *Applications of Digital Image Processing* III, *SPIE*, **207**, 247–267.
[4] Melzer, S. M. (1978). An image transform coding algorithm based on a generalized correlation model. In *Applications of Digital Image Processing, SPIE*, **149**, 205–213.
[5] Gaillat, G. (1983). The design of a parallel processor for image processing on board satellites: an application-oriented approach. *Proc. 10th Annual Symposium on Computer Architecture*, Stockholm, June 13–16, (ACM), pp. 379–386.
[6] Gerritsen, F. A. (1982). Design and implementation of the Delft Image Processor DIP-1. National Aerospace Laboratory NLR Internal Report TR 82083 U, Amsterdam. The Netherlands (also Ph.D. thesis Pattern Recognition Group, Applied Physics Department, Delft University of Technology, Delft, The Netherlands).
[7] Mc Nulty, P. J. (1983). Charged particles cause microelectronics malfunction in space. *Physics Today*, January, **36**, No. 1; 9, 108–109.

Index